WAVERLEY

Paddler for a Pound

The famous £1! Douglas McGowan, director of Waverley Steam Navigation Co., passes the pound note to Colonel Sir Patrick Thomas, chairman of the Scottish Transport Group, at Gourock pier, 8 August 1974. Also in the picture are (left) John Whittle, deputy chairman and chief executive of Caledonian MacBrayne Ltd, and Terry Sylvester, chairman of Waverley Steam Navigation Co. Ltd. (Travel Press & Publicity)

WAVERLEY

Paddler for a Pound

DOUGLAS McGOWAN

Waverley's illustrious predecessor, also named *Waverley*, was built in 1899 for the North British Railway Co. and was tragically lost at Dunkirk in 1940. (Douglas McGowan collection)

To the dedicated officers and crew of PS *Waverley*,
past, present and future

First published 2003
Reprinted 2003

Tempus Publishing Limited
The Mill, Brimscombe Port,
Stroud, Gloucestershire, GL5 2QG

British Library Cataloguing in Publication Data.
A catalogue record for this book is available from the British Library.

ISBN 0 7524 2877 2

Typesetting and origination by Tempus Publishing Limited
Printed in Great Britain by Midway Colour Print, Wiltshire

Contents

Acknowledgements

Many of the photographs are from my own collection, acquired over many years. I am indebted to the following for allowing me to use their material:

John Goss
The Scotsman (www.scotsman.com)
Lawrence Macduff
George Young, George Young Photographers
Second City Films Ltd
The Herald and *Evening Times* © SMG Newspapers Ltd
Ian Quinn
Joe McKendrick
John Innes
John Beveridge
Alistair Deayton
Fraser MacHaffie
Gordon Wi lson
Bill Dempster
Travel Press and Publicity

My special thanks go to John Whittle, former chief executive and deputy chairman of Caledonian MacBrayne Ltd, who so kindly and readily agreed to write the foreword. I am also indebted to my wife Jean for assisting with typing and being a constant source of encouragement (over many years!) and to Joe McKendrick for his patience in proof-reading the text and photograph captions.

A number of photographs are from the superb collections of the late Ian Shannon and the late James Aikman Smith. My thanks are also due to those anonymous photographers of years gone by whose efforts have survived to embellish this publication. If any copyrights have inadvertently been infringed, there has been no deliberate intent, and for anyone whose contribution has not been acknowledged, the fault is entirely mine, as are any factual errors in the text.

Bibliography

Duckworth, C.L.D. and Langmuir, G.E., *Clyde River and other Steamers* (Brown, Son & Ferguson, 1937)
Stephenson, Pamela, *Billy* (HarperCollins Entertainment, 2001)
Williamson, James, *Clyde Passenger Steamers from 1812 to 1901* (MacLehose, 1904)
Waverley – the Golden Jubilee (Waverley Excursions and Allan T. Condie Publications, 1997)

Preface

It is sometimes difficult to believe that the *Waverley* is about to start her twenty-eighth season in preservation and her fifty-fifth season of operation. This is the story, told largely by photographs, many of which have not been seen before, of the part I played in the 1970s helping to return the world's last sea-going paddle steamer to service. It is also the story of the efforts of a small group of enthusiasts rising to a challenge against all the odds. It is the story of determination, disappointment, frustration, enthusiasm, excitement and finally success! It is the fairytale story of the ship famously 'sold' for £1, reincarnated in 1975 to become one of the best known ships in the world, leading a life which was to be much more demanding and spectacular than her first life.

Since 1975, *Waverley* has become one of Scotland's top tourist attractions, and elsewhere in the UK where she operates, huge crowds line the promenades and flock up the piers to witness this phenomenon of a bygone age.

On a warm summer's day, I find it enormously satisfying to witness the *Waverley*, packed with holidaymakers, tourists and locals, all enjoying themselves. What could be more idyllic than sailing up the beautiful Kyles of Bute, or cruising along the spectacular north Devon coast to Ilfracombe or round the Isle of Wight, on a sunny day with the sun reflecting on the water, the hypnotic beat of the paddles thrashing the water, listening to the unique sound of the steam whistle, watching those massive triple-expansion steam engines hard at work, sniffing the intoxicating aroma of the hot oil and steam or simply relaxing on deck taking in the magnificent scenery? Sheer bliss! I believe this is what we have preserved for future generations to enjoy.

Whilst I played my part in helping to return *Waverley* to service in 1975, many others have since demonstrated much more courage and determination. I refer to those individuals both ashore and afloat who have dedicated a huge chunk of their lives to 'keeping the show on the road' despite, at times, appalling adversity. Voluntary board members, officers and crew, support shore-based staff and an army of volunteers around the UK deserve my heartfelt thanks for their unique contribution, often above and beyond the call of duty. Particularly in the formative years, the business would not have survived without the tenacity and determination of both Terry Sylvester and Captain David Neill in particular.

In the winters of 1999/2000 and 2002/2003, *Waverley* underwent two major pieces of surgery. Firstly, the entire aft section of the ship was substantially rebuilt, including two new boilers, new funnels and a major refurbishment of the dining saloon and galley. This has been followed by a major rebuilding of the area forward of the funnels, including major replating of the hull, a new deck shelter, new passenger lounges, a thorough overhaul and refurbishment of the main engine and paddle wheels, and the installation of a new sewage plant. This has only been possible through major grants from the Heritage Lottery Fund of £6 million and the Paddle Steamer Preservation Society. Smaller contributions have come from Glasgow District Council, Scottish Enterprise Glasgow, the European Regional Development Fund and local authorities.

I thank them all for having the foresight to recognize and support a unique maritime preservation project. Their generosity will hopefully allow *Waverley* to sail on for many years.

It is a sad but true fact that in *Waverley* and her consort, the motor ship *Balmoral*, we have the only remaining large-capacity ships in which to view the UK coastline. What a sad indictment for a once proud maritime nation!

Waverley's biggest enemy is complacency – she must generate the maximum revenue each operating day to cover rapidly escalating costs. And so if you have not already had that magical *Waverley* experience, I would encourage you to come aboard soon. A warm welcome awaits you!

Douglas McGowan
Gretton, May 2003

Foreword

We are fortunate today that, thanks largely to enthusiastic volunteers, many historical transport artefacts have been preserved in an active role giving us the opportunity to step back in time and experience the transport of yesteryear.

Prominent among these is PS *Waverley*, unique in that she is the sole surviving sea-going paddle steamer. It is thanks to the efforts of the author and his colleagues that we can still enjoy a trip on this magnificent ship and follow the hallowed tradition of visiting the engine room. While the diesel engines on modern ships hum away in decent obscurity, *Waverley*'s engines are open to view with those gleaming pistons and connecting rods busily driving her through the water.

Paddle steamers have a proud history. They played a vital role in the development of the Clyde and Western Isles networks inherited by Caledonian MacBrayne. Long the mainstay of these networks, they not only provided the essential lifeline links but also gave excellent opportunities for pleasure trips among glorious scenery.

The first challenge to their supremacy came with the development of efficient screw-driven ships then increasing vehicle traffic inevitably caused their basic role to be usurped by the modern car ferries which now tread the paths once followed by the paddlers.

Waverley soldiered on for some years, operating a variety of excursion sailings on the Clyde but these proved increasingly unremunerative. Finally the prospect of expensive boiler repairs prompted her withdrawal after the 1973 season. While such decisions are inevitable on strict commercial grounds, few shipowners or their managers avoid an emotional response. All ships, however humble, generate affection – one reason, perhaps, why they are always referred to as 'she'.

And *Waverley* was a very special case. Cold commercial logic decreed she should be sold for the best price achievable, meaning she would go for scrap. But she was part of our heritage; the last of the line and our thoughts turned to the possibility of her preservation.

Caledonian MacBrayne were not equipped to undertake this. A Clyde Maritime Museum had been rumoured and could have been an ideal home for *Waverley* but failed to materialize. However, we enjoyed an excellent relationship with the Scottish Branch of the Paddle Steamer Preservation Society who had done much to support *Waverley*'s excursion programme and had contributed many helpful suggestions to develop both the operation and marketing. Impressed by their undoubted enthusiasm and ability, the decision was taken to offer them the opportunity to preserve *Waverley*.

It was with some trepidation that I approached Douglas McGowan with this proposal. After all, we were asking them to take on a major responsibility. I am sure Douglas will not dispute my recollection that he reacted with mixed emotions; a combination of joy and consternation would seem the most appropriate description!

Naturally, Douglas had to consult his colleagues and they agreed to take up the offer, presumably after many sleepless nights! A new company, Waverley Steam Navigation Co. Ltd, was formed to own the ship. Colonel Thomas, Chairman of Scottish Transport Group, then owners of Caledonian MacBrayne, and I handed ownership to Terry Sylvester and Douglas at a little ceremony in Greenock. Caledonian MacBrayne's coffers were 'swelled' that day by a token payment of £1!

Preservation as an exhibit at a permanent berth would have been a considerable achievement but, somehow, sufficient funds were raised to cover the substantial costs of returning *Waverley* to service. Since then *Waverley* has gone from strength to strength with regular summer cruises both in her natural home on the Clyde and further south.

Craigendoran pier, 1946. Two LNER paddlers are alongside: the legendary *Lucy Ashton* and diesel-electric vessel *Talisman*, built by A&J Inglis Ltd in 1935. (Douglas McGowan collection)

It is a very real privilege to contribute a few words to this excellent book about *Waverley*'s early days in preservation. Douglas outlines the trials and tribulations and, more importantly, the triumphs of this period culminating in *Waverley* operating her inaugural cruise under new ownership in May 1975.

I am sure the reader will enjoy this story and share my admiration for the determination, tenacity and sheer hard work of those involved. I wish them and *Waverley* continuing good fortune and hope that this fine ship will sail on for many more years to come.

John Whittle

John Whittle
Former Deputy Chairman and Chief Executive
Caledonian MacBrayne Ltd

One
In The Beginning

The colourful history of the Clyde steamers has been extremely well documented over the years and it would be unnecessary and indeed quite improper for me to even attempt to add to the wealth of material which already exists on the subject.

Moreover, the real purpose of this work is to focus on the years 1972 to 1978 when I played a key role in acquiring the *Waverley* from Caledonian MacBrayne Ltd on behalf of the Paddle Steamer Preservation Society, restoring her and returning her to service. Nevertheless, to put the story into context, we do need to go back in history, albeit briefly.

Waverley is the very last of a proud dynasty of Clyde paddle steamers going back to Henry Bell's famous *Comet* of 1812 which was to be the pioneer of over fifty paddle steamers operating on the Clyde in the late 1890s. Such wonderful evocative names as *Redgauntlet*, *Dandie Dinmont*, *Kenilworth*, *Glen Sannox*, *Lucy Ashton*, *Lady Rowena*, *Neptune* and *Lord of the Isles* conjure up colourful recollections of steamers in cut-throat competition racing to be the first at the pier. In the late 1890s, the three railway companies, Caledonian, North British and Glasgow & South Western, practically controlled the Clyde passenger traffic. However, although the undoubted winner was the passenger with a marvellous choice of routes and operators, it was at times uneconomical and wasteful. To quote Captain James Williamson from his book *The Clyde Passenger Steamer* published in 1904, 'As a matter of fact, much of the service is superfluous and needless, the sole advantage derived from the running of three steamers where one would serve being to afford the public a choice of route. One has only to watch the converging of the steamers from the coast town of Gourock on a summer morning between eight o'clock and a quarter past. At that hour, there are in sight no fewer than nine steamers on five days of the week and eleven on Mondays. This flotilla is employed to carry passengers who, in the aggregate, could be accommodated comfortably on board three steamers.' But what a feast for the steamer enthusiast!

The fastest steamers were naturally the most popular with the travelling public and the steamer which reached the pier first was a great advertisement not only for her owner but also for the yard which built her. Fortunately, there were few accidents.

I can still remember my father telling me when I was about eight years old (in 1956!) that I was born about fifty years too late. He was referring of course to my fast-developing passion for the Clyde steamers. I suppose for my particular hobby I did arrive on the scene a little late but I will always be grateful for at least experiencing the twilight years of the 'real' Clyde steamers in the 1950s and '60s: the *Duchess of Hamilton*, *Caledonia*, *Jeanie Deans*, *Saint Columba*, *Queen Mary II*, *Talisman* and the superb *King George V*. Here, I have to confess great difficulty in becoming even moderately excited about the ubiquitous car ferry.

So, what really got me 'hooked' in the first place? I guess it was the family holidays in the 1950s to Whiting Bay on the beautiful Island of Arran which really sowed the seeds. It was not just the holiday experience itself which held a fascination for me but the crossing to and from the island. I can just about remember sailing to Arran in the early 1950s on the turbine *Glen Sannox* and *Marchioness of Graham*. Then, during the holiday, a further treat might be in store – an afternoon cruise to Campbeltown on the *Duchess of Hamilton* or to Pladda on the *Caledonia* or *Waverley*. But the paddle steamers held a special magic – the rhythmic paddlebeats, the hypnotic engines whirling round and the frothy white wake at the stern.

As a six-year old, I was smitten! I had become a steamer nutter! When I was twelve, an uncle gave me a gift of a Junior Subscription (2*s* 6*d* or 12.5p) to the Clyde River Steamer Club. Shortly after, I took my treasured album of Clyde steamer postcards to one of the club meetings, at that time held in the Central Halls in Glasgow's Bath Street. Plucking up a huge amount of courage at the end of the meeting, I proudly presented my album to such worthies as Graham Langmuir, Leo Vogt and George Stromier for scrutiny. To this day, I will never cease to be amazed that they demonstrated such patience and interest from a small boy proffering third-rate postcards which they must have already seen a thousand times.

About the same time, my collection of steamer postcards and photographs was rapidly increasing, assisted by another uncle who was a commercial traveller and visited all the Clyde islands regularly. I eagerly awaited his return from his travels as I knew my collection would be swelled further. I even collected with great enthusiasm autographs of all the steamer captains, mates, chief engineers, pursers and stewards. I am sure I must have them somewhere! School peers were collecting pop and film star autographs – I was collecting autographs of Clyde steamer captains! I also began collecting little plaster-cast models of the steamers, sold for the princely sum of 7*s* 6*d* aboard the steamers. Like me, they are now showing signs of age.

One of the most popular steamers on the Clyde was the 1899 paddler *Waverley*, built for service by the North British Railway Co., later to become the LNER. She was tragically sunk at Dunkirk in May 1940 when returning to England laden with troops from the beaches, resulting in heavy casualties. During 1945, the LNER put the order to replace *Waverley* out to tender, the contract being won by A&J Inglis of Pointhouse, Glasgow. The keel was actually laid in December 1945 but due to a post-war shortage of materials, she was not launched until 2 October 1946. She was named *Waverley* by Lady Matthews, wife of the LNER chairman, and she was to be the very last Clyde paddle steamer built. In January 1947, she was towed downriver to Victoria Harbour, Greenock, for installation of her engines and coal-fired boiler by their builders, Rankin & Blackmore of Greenock. It can be said at this stage that, whilst in some areas of fitting out *Waverley* suffered from a somewhat utilitarian appearance, the quality of her triple-expansion engine was impeccable and this has certainly been proved over the years. Many engineers have told me subsequently that with regular maintenance her engine could run for evermore, and so *Waverley* carried out trials between 2 and 5 June 1947, achieving 18.37 knots.

The choice of paddle propulsion was due to the shallow water around Craigendoran which was to be her base. *Waverley* was 240ft long, gross tonnage of 693 and could carry 1,350 passengers on a Class V certificate. (In 2003, she could carry a maximum of 925 passengers). She was designed as a two-class vessel and had a restaurant, tearoom and bar. She wore the attractive funnel colours of red with black top, separated by a white band. It was to be another twenty-five years before *Waverley* would sail again sporting these colours as in the following year, 1948, came nationalization and with it, by contrast, the rather drab buff funnels with black tops of the Caledonian Steam Packet Co. Ltd. Her captain was John Cameron, DSC, who

The new *Waverley*, dressed overall to celebrate her maiden voyage on 16 June 1947. She is seen here approaching Dunoon. (Douglas McGowan collection)

had been navigating officer of the previous *Waverley* sunk at Dunkirk. Her chief engineer was Bill Summers. Although neither of them knew it in 1947, both were to take a very active interest in *Waverley*'s 'reincarnation' some thirty years later.

Waverley's primary route was the cruise to Lochgoilhead and Arrochar via Loch Goil and Loch Long and the connection with the Loch Lomond steamer at Tarbet, marketed as the Three Lochs Tour, and it was this route which she followed on her maiden voyage on 16 June 1947. She was also regularly used for ferry and commuter duties between Craigendoran, Dunoon and Rothesay. For some twenty years, *Waverley* settled into this routine, and in the 1950s and '60s would appear regularly on the popular Round Bute and Arran via the Kyles excursions. She was to be a willing but rather unremarkable workhorse.

Two

The Seeds are Sown

We now fast-forward to 1966 when I joined the Paddle Steamer Preservation Society (PSPS) as a Life Member. This was an organization primarily based in the south of England with Bristol Channel, Wessex and London branches but no representation north of the border. I found this rather odd as I had always regarded the Clyde as the real home of the paddle steamer, where the *Comet* had started it all in 1812 as the first commercial steamship. However, the Clyde River Steamer Club, founded in 1932, had in many respects served the needs of steamer enthusiasts in the west of Scotland with their monthly meetings in Glasgow and a very innovative series of unusual charter sailings, particularly in the late 1960s, which really captured the imagination of the enthusiast and general public alike.

Three years later, 1969 proved to be a significant year for *Waverley*. By a strange coincidence, the year presented four milestones. Firstly, in July, whilst returning from an evening excursion to Colonsay from Port Ellen (Islay), I met David Neill, who at that time was serving as Second Mate aboard MacBrayne's motor ship *Lochnevis*. It quickly became obvious that we shared a common interest in possibly preserving one of the last Clyde paddlers, *Caledonia*. We both suspected that 1969 was to be her final year of service. Six years later, David was to become captain of *Waverley* under the auspices of Waverley Steam Navigation Co. Ltd.

Secondly, in August in the Argyll hotel, Dunoon, on my twenty-first birthday I was introduced to an ebullient and charismatic Welshman, Terry Sylvester from Barry in South Wales. Terry had witnessed the recent demise of the splendid P&A Campbell paddle steamers *Bristol Queen* and *Cardiff Queen* and had turned his attention to the Clyde paddlers *Caledonia* and *Waverley*. Chemistry plays an important part in any relationship and from that first day we met, our mutual thoughts, ambitions and aspirations seemed to run a parallel course.

Thirdly, I was persuaded to launch a Scottish branch of the PSPS by the then National Secretary, Chris Phillips. The branch was duly launched in December 1969 at a meeting in the Christian Institute in Bothwell Street, Glasgow. I seem to remember that about twenty dedicated souls turned up! The society now has a national membership of over 4,000 from humble beginnings! I was to be secretary, and a chairman and committee were duly appointed, with a programme of monthly meetings to follow. The significance of the establishment of a Scottish branch of the PSPS cannot be overestimated. Without its existence and local activity and lobbying which was to follow, it is unlikely that Caledonian MacBrayne (CalMac) would have made their magnanimous gesture four years later.

The fourth and final milestone in 1969 was the rather courageous attempt to save the *Caledonia* from the breaker's yard. Only two weeks after the formation of the Scottish branch, Iain Hunter (Chairman of the Scottish branch), David Neill and myself created a business plan to buy the *Caledonia* from the Caledonian Steam Packet Co. Ltd (CSP) and operate her ourselves. Following a meeting at the *Glasgow Herald* office with their shipping correspondent, a prominent headline appeared the following day on the front page of this influential newspaper: 'Group plan to buy *Caledonia*'. This publicity not only aroused the interest of several sympathetic Glasgow businessmen, with whom we held a number of constructive meetings, but also encouraged Harrison (Clyde) Ltd, the Glasgow shipping company and parent company of Western Ferries, to come forward and express interest in providing, at the very least, the technical management for the proposed operation. Sadly, some weeks later, it became apparent that our plan would never succeed as the CSP were unwilling to sell any of their fleet to a third party who intended to operate in competition with their services. It should be remembered that in 1970 the CSP were still operating an extensive itinerary of Clyde excursions. It would have been difficult, probably impossible, to devise a commercial operation which did not conflict with their services. However, in many respects, although we didn't realize it at the time, this was a sort of 'dummy run' for what was to follow three years later. Furthermore, although our efforts to preserve *Caledonia* operationally had failed, the brewer Bass Charrington did eventually become interested in the steamer, purchased her and converted her to a floating pub and restaurant, to be moored at Cleopatra's Needle, on the Thames in the heart of London. Sadly, she was destroyed by fire in 1980 and subsequently broken up.

With the demise of *Caledonia*, *Waverley* was to proudly become the last in the line of Clyde paddle steamers and the last sea-going paddle steamer in Europe. Within the slowly flourishing Scottish branch of the PSPS, we were more than aware of this and set out to establish the Waverley Study Group, led by Terry Sylvester and myself. Its purpose was to look at ways of broadening the appeal of the ship, to offer advice and ideas to the CSP management and to overall improve her commercial viability. This was established under the auspices of the PSPS and from 1970 Terry and I started to have regular meetings with John Whittle, at that time general manager of the company, latterly chief executive of CalMac.

At every meeting, John extended towards us a very courteous reception, even if all our suggestions were not implemented! On more than one occasion, I recall the CSP cocktail cabinet being opened and leaving the Gourock HQ feeling somewhat flushed! John was a very busy executive running all the steamer and ferry services on the Clyde. That he chose to listen willingly to a couple of 'steamer nutters' at various meetings over four years is a credit to his patience! We strongly believed that in order for *Waverley* to be successful on the Clyde, she ought to have a completely different colour scheme from the other vessels in the CSP fleet, a separate identity to help her to stand out from the crowd. One idea proposed was to revive the old LNER colour scheme of red, white and black funnels and black paddleboxes. The latter was implemented for the 1972 season. From 1970, the PSPS also demonstrated their support for *Waverley* by chartering the steamer on a regular basis to unusual destinations such as Ormidale, Inveraray and Ardrishaig. However, the branch did not forsake the other Scottish paddler, *Maid of the Loch*, and also chartered her on several occasions.

Around this time, my wife-to-be, Jean Martin, and I did some regular courting at weekends aboard *Waverley*. How fortunate that Jean hailed from Port Glasgow, a mere stone's throw from Gourock and even more fortunate that having been brought up on the banks of the Clyde, she had more than a passing interest in the steamers! Providence!

We both became permanent features aboard *Waverley* each weekend and quickly became good friends with the officers and crew: Captain Hugh Campbell, John Ellis (mate), Tommy Peat (chief engineer), Frank Williams (second engineer), John Brewster (purser), Walter Bowie (assistant purser), Angus Kennedy (chief steward), Vera (tearoom), Pat McGhee (greaser/donkeyman), and 'China' (fireman). 'China' actually achieved a mention in Billy Connolly's recent biography by his wife Pamela Stephenson. Billy was quite a *Waverley* fan and apparently became quite friendly with 'China' in the early 1970s. All the officers and crew were great characters, some of them sadly no longer with us. We were both treated well aboard the ship by the CSP crew. Perhaps it was because they appreciated the efforts being made to 'popularize' their ship. I really don't know but we were always treated like royalty.

There was, however, the odd embarrassing moment. *Waverley*, now based overnight at Gourock instead of Craigendoran, would return there following the disembarkation of her Craigendoran passengers on a Saturday evening. Jean and I would often take advantage on a balmy summer's evening by sailing across to Craigendoran, then back to Gourock when the ship had no passengers on board. Upon disembarking at Gourock, the pantryman, or ship's cook, would occasionally stuff a large greasy parcel wrapped in brown paper into my hands as we went up the gangway. On reaching home, we would carefully unwrap the mysterious bundle with some trepidation. It would usually contain enough bacon and sausages to feed an army for a year. Perhaps I had appeared under-nourished.

Prior to the 1972 season, we had acquired a printout from Lloyd's Register of Shipping in London of all paddle-driven vessels throughout the world. This confirmed that *Waverley* was now the very last sea-going paddle steamer in the world. This would prove to be a very powerful marketing tool in years to come.

When I left school in 1966, I had joined the Inland Revenue in Glasgow as a tax officer, a job I rapidly began to loathe. I really could not see myself allocating tax codes to the general public for the rest of my working life! However, it was probably a better choice than the other job offer I had that year as a commercial apprentice with Alexander Stephen Shipbuilders of Linthouse. Sadly, like so many other Clyde yards before and after, Stephen's shipbuilders were to cease trading only a few months later.

And so it was in 1969 that I joined the world of selling and became a trainee sales 'rep' for Terry's of York, manufacturers of the famous chocolate orange, All Gold and lots of other chocolate delights. I was promoted to my own 'territory' a few months later and given a Vauxhall Viva. Aged twenty, I was one of the youngest salesmen in the company. That year, I was to not only meet my wife-to-be but surprised myself by winning the coveted 'Salesman of the Year' award – a luxury weekend at the Waldorf Hotel in London (for two!) – Jean must have thought she'd won the lottery!

My territory covered the south side of Glasgow, Renfrewshire and south Argyllshire which took me to places like Dunoon, Tarbert, Ardrishaig and Campbeltown every six weeks. Sheer heaven! (but not so in January and February!)

On 15 July 1971, I was returning home from my chocolate travels in Kintyre via Arrochar. By a strange and wonderful coincidence, the timing was to coincide with *Waverley*'s departure from Arrochar at 2.00 p.m. down Loch Long. The ship was well filled and it was a pleasant day with a slight on-shore breeze. It was low tide and *Waverley*'s bow was well below the top of the pier as she very slowly went astern on leaving the pier. To my horror, I was suddenly conscious of the bow 'falling in' against the knuckle of the pier as she glided astern, a combination of slow speed, tide and wind. And then calamity! Her bow collided with the pier, causing the rather fragile pier structure to shake violently. *Waverley*'s mast stays then became entangled in the pier causing immediate tension which resulted in the top section of her foremast fracturing which came crashing down on to the foredeck narrowly missing Fergie Murdoch, the former commodore of the Clyde fleet and well-known captain of *Duchess of Hamilton*. Fergie had come out of retirement and was working as mate on *Waverley* for part of the 1971 season and had been chatting to passengers on the foredeck as the steamer departed. He had a very narrow escape, as had the other passengers in close proximity. But worse was to come. The ship's telegraph system had also been damaged in the collision and there was no communication from the bridge to engine-room. By this time, *Waverley* had picked up speed and was doing some 10 knots astern towards the opposite shore of Loch Long, the engineers down below blissfully unaware there

Waverley arriving at Inveraray on a special PSPS charter sailing in September 1972. Note the black paddleboxes, which were painted thus as a direct result of PSPS lobbying. (John Goss)

was anything wrong! The helmsman was urgently dispatched to the engine-room to ask the engineer to immediately stop the engines. The collision had also resulted in several feet of handrail being damaged in addition to the telegraph chain and broken mast.

I arrived at Gourock in time to see part of the ship's foremast being rather ignominiously wheeled up Gourock pier in a British Rail trolley. Full marks to the CSP: following urgent repairs by Lamont's of Greenock the following day, *Waverley* returned to service on 17 July, Glasgow Fair Saturday, but sailed for the remainder of the 1971 season with her 'stump' foremast, giving her a rather odd appearance.

To celebrate *Waverley*'s twenty-fifth anniversary the following year, Terry Sylvester and I suggested to the CSP that the occasion ought to be commemorated in some way. In a meeting with John Whittle, the idea of a special afternoon cruise with invited guests from the PSPS, CSP and media came to fruition on 19 May 1972. The CSP's own splendid bakery at Gourock produced a magnificent cake in the shape of *Waverley*'s paddlebox, and cut at a special ceremony on board by Captain Angus McEachran. The twenty-five candles were blown out by Terry Sylvester's eldest daughter Sharon. The PSPS were presented with a fine collection of ex-railway company silverware and the PSPS returned the gesture by presenting the steamer with a commemorative brass plaque which was duly unveiled by my (now) fiancée Jean and resulted in considerable media coverage. However, it also demonstrated the close relationship which was building between the CSP and PSPS.

Although we didn't realize it at the time, 1973 was to be *Waverley*'s final season in service with Caledonian MacBrayne (which had been formed as an amalgam of the CSP and David MacBrayne's fleet serving the Western Isles). In this year, she was painted in the new CalMac livery of red funnels with black tops with a yellow circle and red lion rampant on each funnel.

However, prior to entering service at Easter, *Waverley*'s funnels had originally been painted with a broad yellow band separating the red and black. The result was, not to put too fine a point on it, ghastly, and it was changed within a few days to conform to the rest of the fleet. However, the positive outcome of this episode was that CalMac were at least listening to our suggestions to give the ship her own identity: in the previous season, they had painted her paddleboxes black, making it quite obvious she was a paddle steamer.

On 29 September, the PSPS chartered *Waverley* for the last time under CalMac for a special cruise round the Lochs. During the cruise, the Scottish branch committee presented the ship's purser John Brewster with an engraved cigarette lighter, in recognition of his co-operation and support of the PSPS over the years.

The steamer had suffered from boiler problems in the 1972 and 1973 seasons and had lost several days off service. Towards the end of the 1973 season, rumours abounded about *Waverley*'s imminent demise. The turbine steamer *Queen Mary II* had undergone a major recent refurbishment and word on the street was that if one of the cruise vessels had to go, it would be *Waverley*, due to the condition of her boiler and consequent reliability. Also, the cruising market on the Clyde was in decline and foreign holidays were rapidly becoming much more fashionable where wall-to-wall sunshine could be guaranteed. One cruise steamer on the Clyde in 1974 could easily meet demand.

Waverley's final day in service for CalMac was to be Sunday 30 September 1973 when she spent the morning undertaking a special charter for a film crew and the afternoon on a cruise round Bute. She then sailed the next day for the James Watt Dock, Greenock, where she was laid up for the winter. I recall going aboard one cold bleak day early in November. Instead of the usual banging and hammering of the engineers carrying out the normal winter overhaul, there was only an eerie silence. The only sign of life on board was the ship's watchman, who told me that all work had stopped the previous day.

Perhaps this did not come as a total surprise. Fearing the worst, I telephoned John Whittle, by now chief executive of CalMac. Although not committing himself on the ship's imminent withdrawal on the phone, he invited me to attend a meeting with him the following week at his office.

I could scarcely imagine the bombshell he would drop a few days later...

Three
Paddler for a Pound

And so it was at 2.30 p.m. on Thursday 22 November 1973 that I found myself sitting in the reception area of CalMac's Gourock HQ, about to be summoned to the chief executive's office. Several different thoughts were racing through my mind. Was this going to be a polite way of telling me that for all the reasons I was already aware of, they had no alternative but to sell the ship? Would she be scrapped? Would a museum express interest? Perhaps she would disappear overseas. What part could/would/should the PSPS play in all this? Would they perhaps offer the ship to the PSPS for a knockdown price? That last thought in hindsight came pretty close to the final outcome but I doubt if anyone, myself included, could have anticipated a knockdown price equating to £1!

The meeting lasted less than an hour with only the two of us present. John was fairly relaxed; I was nervous and apprehensive. What happened at that short meeting is now history. John announced that due to harsh commercial economics, his company had no alternative but to withdraw the ship from service. They were not in the business of operating elderly paddle steamers, especially ones which needed extensive boiler repairs. Following discussion with the Scottish Transport Group, CalMac's parent company, it had been agreed that in recognition of all the interest and support offered to CalMac since the formation of the Scottish branch in 1969, *Waverley* would be offered as a gift to the PSPS. It was anticipated by CalMac that for the ship to operate again represented a formidable challenge and a better future might be to have her as a floating club for our members. It was also suggested that we might need the guidance and support of the Maritime Trust.

In the foreword to this book, John recalls my reactions as a mixture of joy and trepidation. To this description I would add incredulity and consternation. 'Where's the catch?' I thought. I was of course unable to give John an immediate formal response. I had to report back to an urgently convened Scottish branch committee meeting the following evening.

I left CalMac's office dumbfounded but excited. What would we do with the ship? Could we raise funds? Could we get others interested? John had reminded me that in order to transfer ownership, we would probably have to set up a limited company. Also, we would have to appoint a ship's husband, take on responsibility for insurance, harbour dues and pay a watchman. And that was only for starters! The PSPS had very little money. Although the society already owned the small former River Dart paddle steamer *Kingswear Castle*, we had been struggling for years to generate sufficient interest and funds to return her to operational condition. The *Waverley* was three times her size and could carry (at that time) 1,350 passengers!

Why did CalMac make this unprecedented generous gesture when the ship could probably have raised at lease £30,000 had she been sold abroad or £20,000 as scrap value? Although a few sceptics believed it was a 'convenient' way for them to dispose of a historic ship, I believe that their gesture was utterly genuine and they wished to see the ship preserved. It is worth remembering that without their gesture, it is most unlikely that *Waverley* would have survived in an operational capacity for another thirty years and of course continues to flourish today.

A somewhat bewildered Scottish branch committee acknowledged the offer from CalMac but any formal acceptance would have to come from the society's central committee in London. The following day, my committee colleague and good friend John Beveridge and I sat before a rather disbelieving gathering of society officials in London where a major debate ensued. What strings were attached? Why would CalMac want to do this? They simply could not believe or understand the motivation, and who could blame them! There were more than a few doubting

A very proud moment! John Whittle and Terry Sylvester look on as Douglas McGowan pays the pound to buy the *Waverley* from Colonel Sir Patrick Thomas of the Scottish Transport Group, parent company of the Caledonian Steam Packet Co. Ltd. Only a few minutes earlier, Pat Thomas had removed a pristine pound note from his wallet to give to Douglas McGowan, thus making the 'sale' of the *Waverley* an outright gift! (George Young Photographers)

Thomases at that meeting and perhaps I had to use all my selling skills to win them over. Thanks to the foresight and faith of the society's founder and president, Professor Alan Robinson, and a few others on the committee, the vote was narrowly in favour of accepting the steamer.

Now the fun and games really started! A company was acquired 'off the shelf' and renamed Waverley Steam Navigation Co. Ltd. We would share another friend Peter Reid's Glasgow office as the WSN registered office. Peter would look after the legal and financial responsibilities and became our financial director. Almost immediately, as branch secretary, I organized PSPS 'work parties' each weekend to work on our new 'baby', carrying out tasks which were many and varied: chipping off rust, painting, varnishing, maintaining the ships pumps and auxiliaries, etc. My technical and practical skills are, as my wife will immediately testify, non existent, and so I was quick to appoint work party leaders.

We had professional joiners, plumbers, engineers, salesmen, a sanitary inspector, naval draughtsmen and even a surgeon in James Moore, who dedicated himself to cleaning out the ship's bilges which he often said jokingly was no different from working on human innards! James was to later become chairman of Waverley Excursions Ltd. On an average weekend, we had up to thirty volunteers working conscientiously amidst a great team spirit. Some of the work undertaken at weekends was probably in hindsight cosmetic but psychologically it was a great motivator. Also, I took the opportunity to invite local press, television and radio down to the ship to report on progress. I enjoyed working with the media and found it rewarding when our efforts resulted in a piece of worthwhile exposure and publicity for the ship. However, I was to later discover in1975 the 'Jekyll and Hyde' personality of the media when things went awry.

Late in 1973, I was contacted by Eddie Hirst who managed the information bureau in Glasgow's George Square. He was also secretary of the Clyde Tourist Association, funded by the old Glasgow Corporation and Clyde tourist resorts. The chairman of the CTA was Sir William Gray who had also been appointed the Lord Provost of Glasgow, a useful contact! At a meeting with Eddie, he expressed interest in supporting a plan which might return *Waverley* to service. Up until now, I have to confess to the vision of *Waverley* being preserved statically as a 'stuffed exhibit' somewhere was uppermost in my mind. To actually operate this large ship ourselves just seemed a bridge too far. But it was Terry who helped to paint the picture in my own mind of the ship actually sailing again. This, coupled with the CTA coming forward, provided the inspiration for me to see this vision as a possibility. Surely it was a goal worth chasing. Nothing ventured, nothing gained and all that!

Eddie Hirst and the CTA had well established contacts not only in Glasgow Corporation but also in the Highlands and Islands Development Board and the Scottish Tourist Board. Various meetings followed with all the other tourist and interested bodies, including the Department of Trade and Industry, and independent marine consultants were engaged to assess the ship's condition.

With funds raised though the PSPS and the CTA, the steamer was thoroughly examined. It was known that when she was built in 1947 just after the end of the Second World War, when quality materials were scarce, *Waverley* was somewhat utilitarian in her outfit. We therefore awaited the consultant's report with some apprehension.

In February 1974, proudly flying the PSPS flag, *Waverley* was towed from the James Watt Dock, Greenock, to the slipway of James Lamont at Port Glasgow for underwater hull inspection. The vessel was still in CalMac ownership at this stage so Captain Hugh Campbell was technically in charge of the move upriver. I was aboard for the historic tow, and several journalists and three television film crews met the ship on arrival at the slipway. I had persuaded Eddie Hirst that it would be marvellous if the Lord Provost, Sir William Gray, could be available to 'lend a hand' pulling the ship out of the water. That photograph appeared the next day on the front pages of the *Scotsman* and the *Glasgow Herald*, as well as achieving widespread coverage on Scottish television and radio news bulletins.

The cost of 'slipping' the ship, towage, insurance and survey was almost £4,000. It was gratifying therefore to learn that the ship was basically sound, although a substantial amount of money would be required before she was again able to carry passengers. The surveyor reported that, following ultrasonic testing of the hull, no major replating would be necessary for ten years. A number of boiler tubes would need to be replaced. This gave us great encouragement to increase the momentum in our campaign to return the ship to service.

During the next few months, there were more and more meetings with interested bodies. This was a challenging time. I was employed full time by the chocolate company to deliver certain volume targets each month and they were operating in a very competitive environment. Just occasionally, chocolate got in the way of saving paddle steamers, and vice-versa. Thank goodness I had an understanding area sales manager (or perhaps he didn't know the half of it!).

Although Terry Sylvester was some 400 miles away from Glasgow in Barry, South Wales, that chemistry was now well established between us – our mutual thoughts and ideas always seemed to be on a parallel course and there was seldom any conflict. After all, we had a common objective: to return this two-wheeled monster to service. Perhaps it was because we were both from a selling background, and essentially salesmen must have a positive disposition! But the important thing was that we worked well together although there were a few occasions when Terry arrived at our tenement flat in Glasgow at 6.00 a.m., having just arrived off the overnight sleeper, when I would happily have gone back to sleep. (After all, I'd only been married for just over a year and Jean was expecting our first baby!) I think Jean was beginning to suspect by this time that she had also married the *Waverley* – she was a very patient lass.

As the summer of 1974 approached, it became increasingly obvious that *Waverley* would not sail that year. Whilst all the dialogue continued about her future in Glasgow's City Chambers and elsewhere, her paddles lay idle for the first summer in twenty-eight years.

The last sea-going paddle steamer in the world was formally handed over to Waverley Steam Navigation Co. Ltd on 8 August 1974 at CalMac's Gourock office. I had issued press releases and there was a good turnout of the nation's hacks and film crews. Little did I think then that some of those present would not only become good friends but would end up joining the PSPS and even editing the ship's own newspaper! Such is the magnetism of this unique ship. Colonel Sir Patrick Thomas, chairman of the Scottish Transport Group, and John Whittle represented Caledonian MacBrayne while Terry Sylvester and I represented the PSPS/WSN. The ceremony was fairly brief with short speeches from both parties. The Bill of Sale was duly signed and witnessed and the ship's papers handed over. Yours truly was appointed Ship's Husband. This was certainly an additional but unplanned item for my CV! To make the transaction legal, the ship was 'sold' for £1. I had, of course, made a point of ensuring that I had some spare cash that morning but I needn't have worried. At the precise moment, Sir Patrick opened his wallet, produced a pristine Royal Bank of Scotland £1 note and handed it to me. He announced that he wanted the transfer of ownership to be an outright gift. I responded by handing him back his pound note and so we paid not a brass farthing for this historic ship! In hindsight, I do wish that pound note had been kept for posterity.

The event was duly recorded by the media but several photographers present wanted the scene re-enacted at the James Watt Dock, Greenock, some three miles away where the ship was laid up. The four of us then drove to Greenock where once again the famous £1 note changed hands. '*Waverley* sold for £1', 'Never could £1 have bought so much', and 'Paddler for a Pound' proclaimed the headlines the next day.

Although CalMac had probably thought it unlikely that we would be able to operate the ship, nevertheless there were two important clauses written into the Bill of Sale. Firstly, we could not operate in competition with any of their services and secondly, we should not sell the vessel to a third party (at least not without their permission).

The PSPS had sufficient funds to pay for insurance and a watchman in the short term – the longer term might be a problem, however. A favourable response to a letter I wrote to the Clyde Port Authority resulted in dock dues being waived, at least for a few months.

The most urgent need was for money; a minimum of £50,000 was required to refit the ship. (Ultimately it cost over £60,000.) A public appeal for cash was required. Who should launch the appeal? Ideally a public figure. I wrote to Billy Connolly, who was rapidly becoming well known and had featured aboard the ship in the film *Clydescope* in 1973. No response. I then approached Roddy McMillan who played the role of Para Handy in the BBC television series. I had a delightful phone call from Roddy. He would love to do it but was about to have an eye operation. I then approached Bill Tennent, who for many years had hosted the news and current affairs programme *Here and Now* on Scottish television and a very well known personality. Bill was delighted to help and duly arrived at our Glasgow flat on the morning of 25 October 1974. Unfortunately, my car had been broken into the previous evening and so a rather embarrassed Douglas McGowan drove Bill down to Greenock in his extremely draughty Vauxhall Viva, with a pile of broken glass at his feet! Undaunted, Bill played the part perfectly in his inimitable good humoured style and duly launched the appeal from a press briefing in *Waverley*'s dining saloon. I had arranged with the Royal Mail for a local PO Box (No.1947) to be established for all donations. (It was interesting that a few months later the Royal Mail was to kindly donate a Victorian postbox from Perthshire for posting cards on board the ship.)

George Train, an Edinburgh actuary and life-long enthusiast, had now joined Terry, Peter and myself on the board. George's role was to focus on a viable timetable and roster for the following summer.

Money began rolling in: £10,000 came from an anonymous donor and £11,000 followed rapidly from Glasgow Corporation. The appeal benefited from exposure in the national press and soon the appeal would reach the dizzy heights of £100,000, boosted significantly by a grant of £25,000 from the Scottish Tourist Board and donations from local authorities. This was beyond our wildest dreams and would pay for the ship's major overhaul, critical if we were to sail again.

In the autumn of 1974, I started to approach about fifty different companies seeking their support, either on a 'no profit/cost only' basis or to supply their services or materials free of charge. This brought an encouraging and positive response from the majority of those approached. The publicity manager of White Horse Distillers, Gordon McIntosh, was the first to get in touch. Gordon had persuaded his board to completely refurbish the *Waverley*'s lower bar in return for it being named the 'White Horse Whisky Bar' and White Horse being the pouring brand. We readily agreed. That was in 1974 – twenty-nine years later, Gordon and I are still very good friends and see each other regularly. Other offers of help quickly followed: Weirs Pumps, Harrisons (Clyde), Armitage Shanks, Esso, a flag company, a paint company, and even a major chocolate company (unknowingly) who donated the free time of one of their salesmen.

Naivety can be hurtful. I had assumed that our campaign to return *Waverley* to service would have gained unanimous approval. I could not have been more wrong. Several so-called 'enthusiasts' wrote to the *Glasgow Herald* and other newspapers decrying our efforts to save part of the nation's maritime heritage. Some wrote directly to me pointing out that *Waverley*'s structure was not sufficiently sound to justify long-term preservation. It was all a waste of time and public money. Many suggested that the preservation effort would be better served by supporting the turbine steamer *Queen Mary II*. I had no wish to become embroiled in a *Waverley* versus *Queen Mary II* contest. But the criticism continued unabated, some of it from individuals I had previously counted as friends, right up until 1977 when the renamed *Queen Mary* was withdrawn. For me, it was a lesson in life … and there's nowt so queer as folk!

As it was becoming increasingly obvious that *Waverley* might just sail again, CalMac had tentatively agreed towards the end of 1974 to manage and crew the ship on our behalf, and WSN would undertake the marketing. Bear in mind that the PSPS really had no aspiration to operate commercially, on a daily basis, a major excursion steamer – we were essentially just a bunch of enthusiasts playing the part of a lobby group encouraging others to maintain paddlers in service wherever possible and practical. Early in 1975, CalMac's offer was withdrawn and we were literally 'thrown in at the deep end'! We had to find officers and crew ... and pretty damned quick.

Following the transfer of ownership, work parties at the James Watt Dock increased and quite often could be seen working around the ship on weekdays and not just at weekends. We established during early 1974 that the aft peak tank was leaking and filling with water. I therefore had the dubious task of approaching Greenock Fire Brigade who duly obliged by turning out once a month to pump the ship dry! They told me that they treated it as a special exercise and the men always enjoyed it (their colleagues across the water in Dunoon were also to come to our rescue in a more dramatic fashion some three years later).

Also, immediately following the transfer of ownership, we felt it important to let the world know that *Waverley* had a new identity. Scott's Shiprepairers at Greenock were therefore engaged to repaint the funnels in the original LNER livery, something Terry and I had wanted to see for a long time. A thorough independent examination of the boiler, including a pressure test, led to the conclusion that it would last another season without major surgery. 8,500 linear feet of timber decking (over 1.5 miles) were repaired or replaced and about 250 bags of soot were removed from the boiler tubes.

To avoid any further unwanted criticism, I went to considerable lengths to ensure that not only the proportion of the red, white and black tops of the funnels were correct but also the shade of red used was an accurate representation of 1947. So early in 1975, I visited two of my childhood 'heroes', Graham Langmuir and George Stromier, to have the benefit of their advice on this matter. I think they were both surprised but pleased I had taken the trouble to research this but I did feel it was important. Following the detailed examination of hundreds of photographs from 1947, I briefed John Hamilton, ship repair manager at Scott's. I was horrified, some two weeks later, on one of my many regular visits to the yard, to discover that the white band had been cut too narrow. Back to the drawing board! Stayrings were duly welded on to the funnels between the red and white bands and eventually, the result after painting was quite pleasing. However, it was a source of slight irritation to me for some years thereafter that the result was not an exact replica of the 1947 funnels and the white band was still slightly too broad.

Part of the CalMac Bill of Sale had included a stipulation that we must return to them the four lions from the funnels. Quite what purpose they had in mind for them is something I never discovered. However, as *Waverley* took on her new identity, the four lions were duly burned off the funnels. As the suspension of my modest little Vauxhall Viva only had capacity for two of these extremely heavy beasts, I dutifully returned two of them to Gourock. However, if my memory serves me correctly, the remaining two didn't quite make it back to CalMac HQ – instead they were the star attractions at a PSPS auction in Glasgow which raised much needed funds for *Waverley*. Strangely, there were no questions asked from Gourock. Perhaps someone thought, two funnels, two lions, sounds fine to me!

Four
Waverley Reincarnated

In early 1975, it was becoming obvious that we were in dire need of some engineering expertise. Bill Summers, who had been chief engineer of *Waverley* for twenty-two years, had retired in 1969. I didn't know Bill terribly well and so it was with some trepidation that I thumbed down all the 'W. Summers' in the Glasgow telephone directory. After abortively phoning seven numbers, there was only one remaining. A deep voice which I recognized answered the phone. I introduced myself and explained my mission. 'Aye, I wondered how long it would take before I got a phone call.' Thus started a great friendship with Bill and his wife Anna – they even babysat for us regularly in the months to follow.

Bill played a key role in the coming months not only supervising the engine overhaul but training Ken Blacklock who was to be appointed Second Engineer (and appointed Chief Engineer in 1977).

There were a number of major hurdles to overcome before we could be sure we had a viable business proposition. We had to appoint a captain, officers and crew; we had to appoint a catering company having taken the decision to delegate all catering on board to a third party; we had to construct a commercial timetable which did not compete with Caledonian MacBrayne; we had to appoint a general manager and create a marketing machine; we had to find a suitable base from which to operate the ship; we had to find an office; and we had to dry dock and overhaul the ship and ensure all passenger certificates were issued and deadlines met. But most important of all was to secure a grant from Strathclyde Regional Council to meet an anticipated operating deficit. As Strathclyde was not due to formally come into being following local government reorganization in May 1975 and we urgently needed to dry dock the ship in February, the timelines were not ideal. So with fingers and everything else crossed, we commissioned Scott's Shiprepairers (themselves one of the companies undertaking work for us at favourable rates) to dry dock and overhaul the ship on 17 February, costing £60,000. Fortunately, some three weeks later we received word from Strathclyde that we were being awarded a grant of £30,000. Phew!

In the midst of all this excitement, as the ship was being carefully manoeuvred into the dry dock, the yard manager shouted to me that my wife needed to speak to me urgently. I knew immediately the nature of the problem. She was expecting our first baby and was a few days overdue. A crane quickly lowered a gangway on to *Waverley*'s deck enabling me to disembark from the ship. As I arrived at my car, horror of horrors, I noticed that it had a flat tyre! With no time to change the wheel, a colleague, Ian Burrows of Harrison (Clyde) and a new *Waverley* director, kindly dropped me off at a friend's house at Greenock where Jean was patiently waiting. Following a mad dash to Glasgow's Southern General Hospital, our daughter Lynn was born just a few hours later.

We then placed an advertisement in several newspapers, including the Western Isles area, for officers, crew and a general manager. Captain David Neill, whom I had met back in 1968, had been following our progress with great interest, and his offer to serve as *Waverley*'s master was readily accepted, even although he had no previous experience of handling paddle steamers. I then found myself, along with David, interviewing prospective employees on several days in between selling chocolate oranges and peppermint lumps. To avoid competing with CalMac, we decided to take yet another gamble to base the ship at Glasgow at weekends and Ayr harbour on weekdays, reviving routes which had been abandoned some years previously. We secured the use of a berth at Ayr close to the fish quay and a derelict warehouse at Anderston Quay, adjacent to our berth on the river, was made available by Glasgow District Council for a peppercorn rent and duly converted into office accommodation. We were almost there!

Throughout the early months of 1975, continuous discussions were held with the Department of Trade & Industry's marine and engine surveyors as D-Day became ever closer. The ship remained in Scott's dry dock for several weeks and during March and April there was a flurry of activity as the deadline of 22 May, our 'maiden voyage', approached.

Catering on the Clyde steamers had generally lost money for years. It was clearly a specialist area and we had our hands full already with numerous other challenges! During April, we had approached over twenty catering organizations in central and west Scotland. Only four responded and having looked over the ship, with its rather antiquated coal-fired galley stove, only one company, Commercial Catering of Stirling, actually tendered for the contract. It was Hobson's choice and although our share of the catering profit was miserly, at least catering was now making a small surplus. It was to be the 1977 season before we grasped the nettle and operated all catering on board ourselves. For most years since 1977, catering and the lucrative souvenir shop have made a very important and worthwhile contribution to the overall business.

Each individual board member had their own defined tasks. My own responsibilities for the first three seasons were many and varied! I looked after PR and the media, the souvenir shop (staffing and stocking), the procurement of the ship's flags, designing the company house flag, and the recruitment and staffing of the purser's department.

The local press and television continued to follow the story with great interest and I was regularly interviewed for BBC *Reporting Scotland*, STV and Radio Clyde which, like ourselves, was a fledgling enterprise. I had film crews and radio interviewers turning up on the doorstep of our Mount Florida flat to cover the story (sometimes over breakfast before leaving for the 'day job'!). On one occasion, Jean returned home quite horrified to discover that a BBC film crew from *Current Account* had been to our flat that very afternoon, with reporter John Milne. She was horrified because they had filmed all round the flat, including the bathroom where lifebelts from the *Duchess of Hamilton*, *Caledonia* and *Waverley* were proudly hanging on the walls alongside our twelve-week-old daughter's nappies drying on the pulley (before the days of disposables)!

Newspaper and magazine features were commonplace around this time with my poor wife even ending up in a two-page colour spread in the women's magazine *My Weekly*.

On 10 May, the fires were lit in the stokehold for the first time in twenty months and steam pressure slowly raised over seventy-two hours. Standing on the quayside, looking at the smoke wafting lazily out of those two majestic funnels was a very special moment and represented a real milestone.

A few days later, on a sunny afternoon, the old girl eased herself gently out of the narrow exit of the James Watt Dock into the main channel. There were lots of wheezes and groans coming from the engine which frankly I had never heard before but I suppose any old lady awakened from her slumbers after twenty months would similarly object. Compass adjustments were carried out off Kilcreggan and slowly *Waverley* made her way towards the Cumbraes to the admiring waves and greetings of many onlookers. Standing alongside David Neill on the bridge wing, we both knew we were in business. At least we had the ship almost ready for her first passengers, although none of us had any idea what the future was to hold.

Apart from a few teething problems with some of the engine auxiliaries and water flooding the toilet floors (resulting in us investing in 'duckboards' for the toilets), *Waverley* acquitted herself well on trials and the critical passenger certificates were duly issued.

We decided that the first cruise ought to be a special thank you to all the companies and individuals who had contributed to returning the ship to service. Several civic dignitaries also embarked on Thursday 22 May, including Glasgow's Lord Provost, Sir William Gray, who had so readily demonstrated his support over a year earlier at the Port Glasgow slipway. The fickle Glasgow weather was at its very best and the sun shone as the Strathclyde police pipe band marched proudly on the quay. It seemed as if every journalist and television crew in Scotland had descended upon us and we even had a life-sized 'White Horse' standing between the funnels! Some 350 guests were wined and dined but only a few were aware that Jean and the shop manageress Jean Teviotdale were up until 3 o' clock that morning making the curtains for the ship's dining saloon! It was a proud day for Terry Sylvester and I as the Lord Provost acknowledged our joint contribution to the project in his welcoming speech.

The inaugural cruise was downriver to Dunoon with the ship undertaking a brief cruise along the Cowal coast before returning to pick up our VIP guests. As we passed the remaining Clyde shipyards, we received a wonderful welcome from all the yard workers who found a grand excuse to stop work and give us an almighty cheer and hearty wave. It was an immensely proud day and I had more than a few lumps in my throat as I chatted to our guests and *Waverley* regally threaded her way downriver.

The first smoke! Raising steam for the first time following extensive lay-up at the James Watt Dock, Greenock, May 1975; this was a very poignant moment for all concerned and marked a real milestone. A few days later, *Waverley* would emerge from the dock to carry out trials and compass adjustments. (James Hall Photographers)

Five
Trials and Tribulations

Having got off to a good start with the VIP cruise, everyone was in good spirits two days later when we operated our first public cruise, which was in fact a special charter for the PSPS from Glasgow to Tarbert and Ardrishaig. Again, we were blessed with fair weather and the ship left Gourock with over 700 passengers aboard. Everything went smoothly, including civic receptions at various piers, until the return journey down Loch Fyne from Ardrishaig when there was a loud bang from the starboard paddle wheel. Fortunately, Bill Summers was aboard (as a passenger!) and he supervised temporary repairs. However, our team of engineers were relatively inexperienced and this was certainly a baptism of fire for them. We had broken a paddle float which was eventually removed. We quickly learned that fateful day that the ship could satisfactorily operate with only seven floats instead of eight on each wheel (albeit at slightly reduced speed). We limped into Tarbert some two and a half hours late, eventually arriving at Gourock at midnight. British Rail were at their best on this occasion and had responded to a ship to shore call (no mobile phones in those days!) to operate a special train to take passengers back to Glasgow. Some worthies (including Jean, baby Lynn and myself) stayed aboard for the marathon upriver return to Glasgow where we arrived at 3.55 a.m. Thankfully, this is one of only two occasions when *Waverley* departed on a public cruise on one day and returned on another!

Everyone was mentally and physically exhausted. 'Up the loch without a paddle' declared the bold front-page *Sunday Mail* headline the next morning. Other newspapers irresponsibly reported that the steamer was going round in circles in Loch Fyne all afternoon! I recall Terry and I looking at each other at 4.00 a.m. both wondering what kind of monster we had unleashed. Would we survive until the end of the season? Would we survive until the end of the week? I also recall arriving at Anderston Quay later that morning at 9.00 a.m. quite amazed that the ship was being made ready for her Sunday cruise at the advertised departure time, albeit minus part of the paddle. Thanks to the tenacity and determination of many individuals but particularly the captain and Second Engineer Ken Blacklock, the ship operated until 8 September and carried 121,000 passengers, a most creditable performance in the circumstances.

However, the season did not pass by without further incident and the first few weeks were eventful to say the least! There were frequent problems with the paddles, a mix of inexperience and using the wrong timber for the floats. It soon became clear that frequent tightening of the bolts to secure the wooden floats to the brackets paid dividends. The boiler was also temperamental and gave rise to some problems although overall it acquitted itself well, given its bad press a couple of years earlier.

If 24 May was the longest cruise, the shortest was to follow only a few weeks later. *Waverley* departed her Glasgow berth with a good number of passengers on a beautiful July Sunday morning. Sadly, we got no further than Dalmuir sewage works (the nearest berth we could urgently terminate the cruise), where I had the delightful task of helping to arrange coaches at 11.30 a.m. on a Sunday morning for 750 disgruntled passengers! The surreal sight of a large fleet of Glasgow Corporation double-deckers at the sewage works (a somewhat unusual destination for them) with the 'pong of the Clyde' pervading the atmosphere is one of those indelible memories. *Waverley* had suffered from major boiler failure but it was an incident which ought never to have happened. Once again, we were to pay the price of inexperience.

Later that day, she was towed rather ignominiously back upriver to her Glasgow berth where repairs were carried out for over a week. This was a major setback as the ship was out of service and not earning at the peak holiday period (the Glasgow Fair).

After only a few days of operation, our chief officer (mate) had to depart urgently for personal reasons. This left us in a real predicament as we could not sail without a mate. John McCallum, mate of MacBrayne's magnificent *King George V* for many years, had retired along with the ship the previous year. John had been friendly with my wife and me for a number of years and his inimitable West Highland wit was renowned. It was John who, on our acquiring the *Waverley*, while cruising down the Sound of Mull aboard the *George*, announced drily, 'I see you've bought that load of scrap for £1.' 'Yes,' I replied, not quite sure what was coming next. Silence for a few long seconds. 'Well, you've been robbed.'

And so it was that one evening in early June, I found myself knocking at John's tenement flat in Glasgow's Argyle Street. I explained our dilemma. Would he do me a big favour and help us out for a few days until we found a permanent replacement? With some reluctance, John agreed to 'have a look over the ship' and sail the next day. He had never worked on paddlers, nor, one has to admit, were they his first love.

As it turned out, John stayed with us until the end of the season and his dry sense of humour proved to be a good foil for David Neill in his first challenging season. John was however unused to the manoeuvrability of paddle steamers, compared with the screw-driven variety. On one early occasion, entering Ayr harbour (at a 'fair lick' in order to maintain steerage), John was heard to shout from the port wing of the bridge, 'My God, we'll all be drowned!' But no-one was drowned.

In July 1975, we briefly ran aground on approaching Rothesay pier but a local pleasure boat successfully pulled us off and there was no damage to the ship. We experienced a crew mutiny at Dunoon when the crew threatened to walk off. Many other incidents coloured our first season and certainly kept us all on our toes. However, we were determined to overcome every single obstacle and the show went on. Being responsible for media and PR, I soon found that those nice journalists who had demonstrated all that goodwill were again showing signs of their split personality when things occasionally didn't quite go to plan. Naivety was again showing itself and I was soon to learn that *anything* which sold newspapers would attract coverage and if it was bad news, so much the better.

The year 1975, however, was not without its humour. I had thought it a good idea to have a piano on board the ship to entertain the passengers. I had contacted William Tennent whose trio was the last of the traditional Clyde steamer bands. He had played aboard numerous steamers including the *Duchess of Montrose*, *Duchess of Hamilton* and *Queen Mary II* and had readily agreed to perform aboard *Waverley* to revive the tradition. When my wife's Auntie Mary's piano became surplus to requirements, Scott's Shiprepairers had duly dispatched their lorry to collect the piano from Auntie Mary's house in Greenock. The piano was put aboard the ship in the aft deck shelter and lashed down with heavy ropes, just in case it went walkabout in a Force 8 gale. However, there was one major problem. Pianos do not like salt water, or even

The ship's officers and crew, pictured at Rothesay pier, September 1975. Despite numerous setbacks during the first season, *Waverley* quickly established herself as a major tourist attraction. (Douglas McGowan collection)

salt in the atmosphere. Especially Auntie Mary's as, like its owner, it was not in its first flush of youth. One of the many PSPS volunteers was Tom Craig, a piano tuner by profession. After retuning the piano on no less than four occasions over a few weeks, Tom sadly had to declare Auntie Mary's piano a lost cause.

A few days later, while *Waverley* was passing Garroch Head, an executive decision was made to jettison the piano overboard, to the utter bewilderment of the onlooking passengers. It was expected to sink to the bottom of the Firth of Clyde. It didn't. It floated. Minutes later, *Waverley* took a radio call from the skipper of a nearby fishing boat, very concerned as he had seen a large object fall overboard from the paddler and it had appeared on his radar screen. His response on being advised that it was Auntie Mary's piano cannot be recorded here. The moral of this story? Pianos float!

One of our deck crew in our first season, old Alec McLeod from the Island of Lewis, had an uncanny knack of catching seagulls from the aft promenade deck, tempting them with bread or other titbits. He would then carry the screeching bird down below, walk through the packed dining saloon, much to the astonishment of watching passengers, and disappear into the galley where he would quickly release the terrified bird through the galley porthole. He would then return to the dining saloon, clutching a roast chicken dinner, announcing to all the passengers 'Anyone for fresh breast of chicken then?' Quite what the RSPB would make of this I have no idea. No wonder seagulls no longer follow the *Waverley*.

John McCallum's unique wit also kept us going during a tough season. On one occasion, a very 'perjink' (posh) lady was having trouble coaxing her little Pekinese dog up the gangway at Dunoon. John was becoming impatient as we were already five minutes past our advertised departure time. The dog steadfastly refused to mount the gangway. 'You'd be better fitting wheels to it,' John suggested. The lady was not amused, glowered at John and picked up the precious animal in her arms.

Reg and Bunty Collinson were our shopkeepers, assisted by Jean Teviotdale. Bunty was a tireless volunteer, working long hours each day for the *Waverley*. She was fiercely proud of what had been achieved and would never hear a critical word said against the ship. One August evening, on a Radio Clyde Showboat cruise, the steamer was packed with beer-swilling youngsters, enjoying the music and dancing on the deck. But it wasn't just alcohol which was flowing that night, the testosterone also seemed to be in good measure as a rather tipsy youth asked Bunty if she stocked contraceptives. (Remember, this was 1976!) The blood drained from Bunty's face and a look of horror mixed with sheer disgust took over. 'No dear, we certainly don't stock *that* sort of thing on *this* ship.' End of conversation.

Despite the trials and tribulations of the first season and considering there was no infrastructure in place and a total lack of experience in operating a large paddle steamer, it was generally agreed that we had acquitted ourselves well. In particular, David Neill and Terry Sylvester must take the credit for providing the essential motivation and leadership, especially when the chips were down.

It is sometimes difficult to believe that in early 1975 everything was set up from scratch: we had no head office, no workshop or store, no network of agents, no fuel supplier, no souvenir suppliers, no tickets, no catering company, no robust relationships established with the DTI, pier owners or advertisers. WSN Director Peter Reid said on the BBC *Current Account* programme, 'We started with faith, hope and charity and we continue to run with faith, hope and charity.' Very true. And we also had the amazing support of the general public who regularly flocked up the gangways and a loyal volunteer network. It was very much a team effort, all striving towards a common objective and it was working, against all the odds.

Six
Southern Safari ... and a Mishap!

As we looked forward with eager anticipation to the 1976 season, it came as a shattering blow to learn in February of that year that Strathclyde Regional Council had decided to award the total cruising grant available (£67,000) to CalMac and to give nothing to WSN. Our finances were already in a perilous state and this decision by Strathclyde, almost certainly politically motivated, was virtually the final nail in the coffin. A last ditch attempt was made to avoid liquidation and a second public appeal was launched which astounded us by raising £15,000 in ten days. The *Sunday Mail* newspaper also launched a major appeal and must take much of the credit for raising public awareness and for throwing us a lifeline. The Scottish Tourist Board contributed a further £15,000, having donated £25,000 in 1975. By May 1976, the appeal had raised over £70,000, ensuring survival for another season.

Having abandoned the 'Doon the Watter' sailings from the Broomielaw in 1969, we were surprised to learn that CalMac were basing their own cruise steamer, renamed *Queen Mary*, on our doorstep at Glasgow at weekends to compete head-on with *Waverley*! On Sundays, *Queen Mary* departed at 11.00 a.m. with *Waverley*'s departure timed at the unearthly hour of 9.30 a.m. It was the norm to see Waverley depart with several hundred passengers and *Queen Mary* with only a handful on board. It appeared that we had little to worry about.

Fortunately, the weather in 1976 was even better than 1975 with long hot summer days and one of the main challenges of our second season was to deal with an unprecedented volume of business. We were forced to make urgent improvements to the ship's passenger capacity and one of the wooden lifeboats was removed from the aft boat deck to be replaced by inflatable liferafts. This had the result of increasing our Class III passenger certificate, critically important for the successful market from Ayr. In addition, the Purser's Office acquired a second ticket window to reduce the long queues, especially on the relatively short Ayr to Brodick route.

Mechanical performance was improved in 1976 partly due to the boiler being retubed at a cost of £20,000 and partly due to us being on a sharp learning curve! We were still encountering problems with the paddles, something that was not eradicated until the 1977 season.

On one busy August Saturday in 1976, *Waverley* was returning to Glasgow when the crucial No.1 float on the port wheel failed off Langbank. Rather embarrassingly, only half an hour earlier, we had overtaken our old rival *Queen Mary* as she was also returning her passengers up river to Glasgow. Limping in to Old Kilpatrick oil terminal, our passengers were rather sheepishly transferred to the *Queen Mary*. It was generous of the CalMac vessel to come to our rescue although perhaps there was just a hint of smug satisfaction in her wheelhouse that evening as she came alongside! Our passengers were returned safely to Glasgow aboard *Queen Mary*, with all but a stalwart few *Waverley* enthusiasts who refused point blank to transfer to the *Mary* and got the train home instead. (I have to confess that at the end of a long day my wife, baby daughter and I were more than happy to enjoy the regal splendour of the *Mary*!)

We completed the 1976 season successfully, having extended the time-table until the middle of September. Passenger numbers and financial results were a considerable improvement on our first season having carried 191,000 passengers. Confidence was gradually building but one thing was rapidly becoming apparent: the Clyde could only realistically support a ten-week season; May and September were very difficult to market. Yet it was impractical to operate the ship for only ten or twelve weeks – the finances just did not make any sense. There was only one solution – we would have to explore other markets. Largely inspired by Captain David Neill and an invitation from Llandudno in North Wales for *Waverley* to be the main attraction at the forthcoming celebrations of the Pier's centenary, the board with a few reservations, authorised *Waverley*'s first-ever excursion into southern waters.

Given that the ship was built for the sheltered waters of Loch Long and Loch Goil, it was with more than a little apprehension and trepidation that we cast off at midnight from Campbeltown on 28 April 1977, windows boarded up and hatches well and truly battened down, for Liverpool. She arrived there at 4.30 p.m. the following day, having encountered a heavy swell off the Isle of Man but none the worse for her adventure. The next day, she was open to the general public and I had the pleasure of welcoming on board the Mayor of Liverpool at a special evening reception.

Thousands of onlookers line the shore at Fleetwood to welcome *Waverley* on her first visit to Lancashire in May 1977. The ship carried a full complement of passengers on this special excursion from Liverpool. (Ian Shannon)

A crowd of almost 4,000, many in Victorian costume, greeted her arrival at Llandudno on 1 May in glorious weather and for the following week, she undertook a variety of excursions from Liverpool. On the Sunday she undertook a full-day sailing from Liverpool to Fleetwood, passing Southport and Blackpool. The crowds on this occasion were even more amazing. I was standing on the foreshore at Fleetwood awaiting her arrival and the crowd was estimated at over 7,000. I could not believe it. The Beatles could not have been a bigger attraction that day!

I had circulated the Mersey media with various press releases not really knowing what the response might be, but the coverage we achieved was incredible with headlines in the *Liverpool Echo* like 'Steam romance returns to the Mersey'.

Waverley returned to the Clyde having taken a bold step into the unknown, literally testing the water, but it was also to be the basis for what was to follow in future seasons.

If the inaugural cruise of 22 May was to be the most memorable *Waverley* day for me personally, then 15 July 1977 was a close second but for all the wrong reasons. With the euphoria of the innovative success of our Mersey excursion behind us, we settled down into the routine of a third Clyde season. What was to follow on Friday 15 July was anything but routine. The following day was the start of the Glasgow Fair holidays, the 'icing on the cake' when traditionally the steamer had carried good numbers and we had a chance to recover the usual deficit incurred in the early weeks of the season. I was aboard, assisting in the ship's souvenir shop, as we returned down Loch Long to Dunoon to board our passengers for the cruise back to Glasgow. With 630 passengers aboard, Captain Neill discovered to his horror that the ship was not responding to the starboard helm as anticipated on the approach to Dunoon. (This was possibly aggravated by the considerable list to port caused by the majority of passengers waiting to disembark.) It rapidly became obvious that *Waverley* was bearing down on the red can buoy north of the Gantock rocks. To avoid damaging the port paddle wheel, emergency full astern was rung on the engine room telegraph and soon she was making good speed astern. However, the fast ebbing tide had the effect of pushing her southwards (the ship's ability to steer accurately whilst going astern in certain wind conditions had been a source of some concern since 1947), and she failed to clear the Gantock rocks. There was a thundering crash as her stern bounced on to the end of the rocks. A few passengers were momentarily knocked off their feet by the impact but there was no panic on board and thankfully no one was injured. Both funnels were seen to visibly rock on impact and the funnel stays parted. I had been serving a long queue of passengers in the shop when I heard the engine room telegraph's unfamiliar emergency double ring and shortly after came the impact.

An immediate attempt was made to free her by asking all passengers to move towards the bow but with the fast ebbing tide, she was stuck fast. The CalMac ferry *Juno* stood by for an hour until all passengers were landed safely by Western Ferries *Sound of Shuna*. Some passengers were also landed by *Waverley*'s lifeboats and the pilot cutter *Gantock*. Once all the passengers were safely put ashore and over an hour later, two tugs arrived from Greenock. But it was too late. My pride and joy was now settling down on to the rocks, accompanied by the sickening sound of steam pipes bursting and steel plates buckling under the strain down below. After two abortive attempts at pulling her off during which one of the tugs' steel hawser pulled the 'bits' out of *Waverley*'s foredeck, the officers and crew set about the urgent task of shoring up bulkheads and preparing the pumps to hopefully contain the water which was already beginning to pour into the lower bar and coffee lounge (which was to be renamed the following season in a touch of typical *Waverley* self-deprecation as the 'High and Dry Bar'!).

During this time, the media who had followed our trials and tribulations so closely were now flying above our heads in helicopters and small planes. This was one major *Waverley* event I was wishing they had not covered. On the ship's bridge, I dictated a press statement via the archaic ship to shore one-way radio to the effect that the ship would be back in service within a week. Optimist to the last! (I was only five weeks out!)

The arrival of the US Navy from the Polaris base at the Holy Loch was a blessing. What an incredible coincidence that it was the same group of people who had chartered the *Waverley* that very same evening for a downriver cruise. The US Navy brought on board heavy-duty pumps which were quickly rigged up and helped to keep water levels from rising any further.

By 6.00 p.m. there was little else I could do and I was relieved to remove my lifejacket and accept a lift back to shore with a press photographer who had shown initiative by hiring a small motor launch and boarding the ship (totally uninvited!). Back at Dunoon, I was interviewed by a television news team which was broadcast later that evening. One of the many phone calls which followed was from my area sales manager, rather bemused at seeing me on television when I was supposed to be selling chocolate in Aberdeen that day! Oops!

Just before midnight, my phone rang at home. It was the ship's purser phoning from Dunoon to tell me that the ship had floated off on the high tide and under her own power had tied up at Dunoon's coal pier. There is little doubt in my mind that had it not been for the quick action of the officers and crew in safely and professionally evacuating all passengers and doing everything possible to save the ship thereafter, the outcome could have been quite different. Many marine experts were confounded that the ship did not break her back, like several other vessels before her which had been unlucky enough to encounter the Gantocks. Some said that the fact that she was a paddler, with the strength of the triple-expansion engine amidships, helped to save her. One thing is for sure, she would not have survived perched on those rocks for many more tides.

The next day, I returned to Dunoon. She was a forlorn sight, sitting on the sandy bottom in the unfamiliar setting of the coal pier with the Dunoon fire brigade in attendance pumping water, the second occasion that the local fire brigade had come to our rescue. Attempts were made to seal up the gashes in the hull with quick-drying concrete. As I inspected the after end of the promenade deck, it was strangely undulating, the normally level wooden planks badly buckled by the grounding. Down below, it was heartbreaking. Pumps were everywhere and the bar and coffee lounges unrecognisable. Stanchions supporting the deckheads were badly twisted and there was devastation everywhere. As I set about the task of de-stocking the shop and loading everything into my car, the same ghouls who had wished us nothing but bad luck three years earlier were to be seen raking over the ship, armed with cameras to record the event for posterity. Perhaps they thought it would be their final photographic opportunity.

The DTI eventually gave us permission to sail five days later to Scott's Garvel dry dock for a hull inspection and strengthening prior to permanent repairs (if that was deemed to be possible).

Naturally, this tragedy was a real body blow to all our aspirations and frankly, no one really knew if the short voyage from Dunoon to Greenock would prove to be her last. Just in case, those on board were determined to go out in a 'blaze of glory' or, alternatively, demonstrate to the world that the old *Waverley* was far from finished. So, in the afternoon of 20 July, she sped across the Firth with the engines flat out, the engineers giving it everything. Her speed was estimated to be at least 17 knots and the bow wave she created was the evidence.

A week later, *Waverley* was moved to Govan dry dock for permanent repairs, costing £78,000. This was a tidy sum in 1977 and as the ship's hull and machinery were insured for only £80,000, she only narrowly escaped being declared, in nautical terms, a constructive total loss by our insurers. *Waverley* was to be out of service for six weeks of the peak season.

We had a major dilemma: *Waverley* was our sole source of income and we certainly could not survive for six weeks without cash flow. It is remarkable but true that throughout *Waverley*'s preservation and especially in the 1970s, when the prognosis often looked bleak, there has always been a miraculous recovery. Against all the odds, Bill McAlpine of the well-known construction company, and a keen steam enthusiast, offered us the charter of his small motor vessel *Queen of Scots* while *Waverley* was out of service. Although much smaller than *Waverley* and lacking her many amenities, it enabled us not only to keep faith with the public but provided much needed cash flow to keep us afloat.

Waverley returned to service on 1 September 1977 with much of her after hull replated following major repairs. We ended the season once again teetering on the brink of liquidation with losses amounting to £77,000. This was only avoided by a moratorium being established with our major creditors. Such was the measure of goodwill for the ship that in the challenging financial trading climate of the late 1970s when inflation was at an all-time high, that several important suppliers were willing to wait patiently for their money.

Having 'market-tested' business on the Mersey in 1977, our appetites had been whetted to venture further south and on Saturday 15 April 1978, *Waverley* departed from Stobcross Quay, Glasgow, and rounded Land's End the following evening to begin the most ambitious and adventurous month in her thirty-year history, masterminded in the main by Captain Neill and Terry Sylvester. Places were visited which five years earlier would have seemed like a pipe-dream: London, Newhaven, Worthing, Ryde, Hastings, Deal, Clacton, Southend, Sheerness, Tilbury, Eastbourne, Portsmouth, Southampton, Bournemouth, Poole and Weymouth. In four weeks she carried 54,000 passengers and had a perfect mechanical track record.

Also in 1978, *Waverley* celebrated the centenary of the famous MacBrayne paddle steamer *Columba* by attempting to match her timings and speed between Ardrishaig and Tarbert, part of the famous 'Royal Route'. She acquitted herself well.

At the end of the 1978 season, when the full trauma of the Gantocks episode had passed and the business had stabilized and was on a sounder financial footing, I decided to step down as a director of WSN. Although my employer had at all times demonstrated full support for my maritime interests, I felt it was 'payback time'. My career was approaching a crossroads: I was ambitious and knew that if I was to progress, I really had to give 110% to the 'hand that feedeth'. This resulted in my appointment in 1978 to Area Sales Manager followed by a number of other moves which meant that the McGowan family relocated south of the Border in 1983 to Cambridgeshire, followed by Berkshire in 1986 and Gloucestershire in 1994. Apart from a brief two-year spell in the 1990s as chairman of Waverley Excursions, my role nowadays is very much on the periphery. I am full of admiration for the volunteer directors of Waverley Excursions and Waverley Steam Navigation Co. who so often go above and beyond the call of duty.

The 1978 venture south thus paved the way for the following years, with the Bristol Channel being added to the itinerary in 1979 and *Waverley* now regularly calls during May and June at Clevedon, Penarth, Newport, Swansea, Minehead, Ilfracombe and Lundy Island. By keeping the paddles turning for almost seven months of the year, the economics of running an elderly paddle steamer in the twenty-first century become marginally more palatable.

* * *

In the past three years, *Waverley* has had over £7 million of Heritage Lottery Fund, PSPS and other public money spent on her. She emerges from the final stage of rebuild from the yard of George Prior at Great Yarmouth in June 2003, probably in better condition than she was when built in 1947. For me, it has been enormously gratifying to see such huge sums of public money going to the ship, twenty-five years after we had to fight for every penny. It is also pleasing to see HLF support going to our other fine ship, *Waverley*'s consort MV *Balmoral*, to enable her to be re-engined for the 2003 season.

Such financial support was never even dreamed of in 1975 and if someone had told me that our plucky little paddler would be sailing into the twenty-first century, I would have been very sceptical, to say the least! But, make no mistake, without that vital injection of HLF cash, there would be no *Waverley* or *Balmoral* today.

I am often asked the question 'How long can *Waverley* survive?' I have always passionately believed that we have in *Waverley* a piece of living history which becomes even more precious and unique with the passage of time – a working dinosaur. She is there for future generations to experience and enjoy, but she will continue to need three vital ingredients: skilled and dedicated officers and crew, the continuing support of the great British public and, last but not least, money.

This is the story of success in the face of adversity. I am very proud to have played a part.

PS *Duchess of Rothesay* was built in 1895 by J&G Thomson of Clydebank for the Caledonian Railway Co. and had a distinguished service record during the First World War. In October 1939, she entered a second period of war service as a minesweeper on the Clyde and then in Dover. She was sold for scrapping in August 1946. (Douglas McGowan collection)

PS *Marmion* entered service in June 1906 for the North British Railway Co. and was used extensively on the Lochgoilhead and Arrochar service. She was designed for winter as well as summer work and was a popular steamer throughout her career. Like *Duchess of Rothesay*, she had an outstanding war record, being attached to the Dover patrol on minesweeping duties from 1915 to 1919. She was again requisitioned by the Admiralty in 1939 but was sunk by an air attack in Harwich harbour in 1941. (Douglas McGowan collection)

PS *Mars* was built by the famous John Brown yard at Clydebank in 1902 for the Glasgow & South Western Railway Co. Sadly, she did not reappear for Clyde service following the First World War as she was hit by a destroyer in November 1918 whilst minesweeping around the approaches to Harwich harbour. An attempt was made to salvage her but she broke her back and became a total loss. (Douglas McGowan collection)

PS *Iona III* was another product of J&G Thomson of Clydebank in 1864. Along with *Columba*, *Iona* had a very long and distinguished career starting with the Ardrishaig summer service from Glasgow's Broomielaw. She remained on the Clyde during the war years but for most of the latter part of her career, she operated in West Highland waters on the Oban to Fort William service. She was sold to Arnott, Young & Co. (Shipbreakers) at Dalmuir in March 1936. (Douglas McGowan collection)

PS *Jeanie Deans* approaching Dunoon pier, July 1938. The *Jeanie* was another Clyde favourite, built by the Fairfield Shipbuilding & Engineering Co. of Govan in 1931 for the London & North Eastern Railway Co. She was much larger than any other previous steamer operating from Craigendoran, being 250ft long. She had a triple-expansion engine, the first of this type built for Clyde service. Following war service when she was placed on minesweeping duties at both Irvine on the Clyde and then Milford Haven, she settled down to a variety of Clyde excursions from Craigendoran. Following a courageous attempt by a group of London enthusiasts to operate her on the Thames as *Queen of the South* in 1966 and 1967, she was sadly towed to Antwerp for scrapping in December 1967. (Douglas McGowan collection)

Opposite above: PS *Kenilworth* was built in 1898 for the North British Railway Co. and achieved a speed of 18.6 knots on trials. She was employed mainly on the Craigendoran to Rothesay route. Following war service and minesweeping duties, first at Troon and later at Portsmouth harbour, she returned to her original Clyde route. She was broken up on the Clyde in 1938. (Douglas McGowan collection)

Opposite below: PS *Lucy Ashton*, one of the most popular Clyde steamers with the travelling public, was built, unusually, by the yard of T.B. Seath in Rutherglen in 1888. Although initially placed on the Holy Loch route, she became closely associated with the Gareloch service, operating from Craigendoran to Helensburgh and villages such as Row, Rosneath, Clynder, Shandon and Garelochhead. She was not 'called up' for either of the two world wars and so faithfully remained on Clyde service throughout. She was finally withdrawn from service in 1949, one of the real workhorses of the Clyde fleet for over fifty years. (Douglas McGowan collection)

PS *Kylemore*. Although launched in April 1897, it was not until 1908 that she operated for the Williamson Buchanan fleet, having been previously owned by the Hastings, St Leonards-on-Sea & Eastbourne Steamship Co. who had renamed her *Britannia*. She sailed for many years on the Broomielaw to Rothesay service and was requisitioned for minesweeping duties in the English Channel from 1915 until 1919. She passed into the fleet of the Caledonian Steam Packet Co. in 1935. She was sunk by enemy action off Harwich in August 1940. (Douglas McGowan collection)

Opposite above: PS *Lord of the Isles* is seen here approaching Dunoon in August 1927, her penultimate year in service. She was launched by D&W Henderson & Co. on the Clyde in 1891. She was always popular with the public and was used originally on the Inveraray service and latterly, following acquisition by Turbine Steamers Ltd, for excursions round the island of Bute from Glasgow. She was the very last Clyde steamer to have polished copper steam-pipes, a unique feature which she retained until the end of her career in 1928. (Douglas McGowan collection)

Opposite below: PS *Jupiter* leaving Dunoon in July 1938. She was built by the famous Fairfield yard at Govan and entered service in 1937. Stowage space for about six cars was provided between the funnels. Like her sister ship *Juno*, she was interesting in that she was built with 'camouflaged' paddle boxes to give the impression of being a more modern, streamlined screw steamer. Following war service as a minesweeper at Milford Haven, Dover and on the Tyne and Humber, she settled down to Clyde service, mainly on the Gourock and Wemyss Bay to Rothesay run. She was rather prematurely withdrawn in 1957. (Douglas McGowan collection)

PS *Mercury* approaching Dunoon pier in June 1938 (with TS *Duchess of Montrose* departing). *Mercury* was built for the LMS Railway in 1934, along with her sister ship, *Caledonia*. Both paddlers, like *Juno* and *Jupiter* some three years later, were built without the traditional style of paddleboxes, but nevertheless had an attractive appearance. *Mercury* was used principally on the Kyles of Bute service from Greenock and Gourock. She was damaged by a mine in December 1940 and sank whilst under tow between Milford Haven and the Irish coast. (Douglas McGowan collection)

PS *Grenadier* pictured here at Oban was built in 1895 for David MacBrayne Ltd by J&G Thomson. She was an attractive vessel, being a saloon steamer with clipper bow and bowsprit, and was used mainly for the summer excursions to Staffa and Iona from Oban. Although not known for her speed, she was nevertheless a good sea boat. She came to an untimely end when fire broke out on board when she was alongside at Oban in September 1927. Sadly, a number of crew lost their lives in the fire and she was later broken up at Ardrossan. (Douglas McGowan collection)

The spark which was to kindle a lifetime's love of the Clyde steamers: *Duchess of Hamilton* alongside Whiting Bay pier on the island of Arran, July 1960. (John Innes)

Whiting Bay pier again, this time with PS *Jeanie Deans* alongside in August 1961, ready to depart for her afternoon cruise to Pladda. Today, at this pretty Arran village, there is no sign of the remains of the pier which was the longest on the Clyde. It was demolished in the early 1960s. (John Innes)

Waverley fitting out at the East India Harbour, Greenock, January 1947. She has been towed downriver from her builders, A&J Inglis Ltd of Pointhouse, to Greenock, where her engines are now being installed. (Douglas McGowan collection)

Waverley's triple-expansion engine on the test-bed at the builder, Rankin & Blackmore Ltd of Greenock, 1946. This engine (2100hp) has certainly stood the test of time and is still the powerhouse of *Waverley* today, some fifty-six years on. (Douglas McGowan collection)

Waverley at Craigendoran in her first year of service, June 1947. The bow of *Lucy Ashton* can just be seen on the left of the photograph. (Douglas McGowan collection)

Waverley laid up for the winter in January 1948 at Bowling harbour with *Lucy Ashton* alongside. *Waverley* sailed with the attractive LNER colour scheme of red, white and black funnels for only her first season, before adopting the rather more drab buff and black colour scheme of nationalization. It would be another twenty-seven years before her original funnel colours would return. (Douglas McGowan collection)

Waverley in Rothesay bay, June 1947. In the background can be seen a large naval vessel at anchor. Rothesay bay was regularly used by the Navy until the 1970s for visiting naval craft. Note the condition of *Waverley*'s hull, particularly the after section, after only a few days in service!

Leaving Kilcreggan, August 1947: returning to Craigendoran from an afternoon excursion. (Douglas McGowan collection)

Leaving Dunoon in July 1955. The double gold band along the black portion of the hull has been retained into Caledonian Steam Packet Co. ownership and was reintroduced by Waverley Excursions as a permanent feature after rebuild in 2000. (Douglas McGowan collection)

Pictured off Gourock, August 1964. This would be the final season with black hull, prior to the introduction in the following year of 'monastral' blue hull with red lion rampant on the funnels. The gold lines along the hull have long disappeared and she now has white paddleboxes. (Douglas McGowan collection)

CALEDONIA

A previous attempt at preservation had also ended in failure: PS *Queen of the South* (formerly *Jeanie Deans*) at speed on the Thames near Greenwich, July 1966. Despite a considerable amount of money being expended refitting the vessel, she only sailed on a few occasions in both 1966 and 1967, being plagued by boiler and paddle problems. (Fraser MacHaffie)

Opposite: PS *Caledonia* berthed at Stranraer harbour on a special charter by the Coastal Cruising Association, May 1965. *Caledonia* was to be the subject of the first serious attempt to preserve a Clyde paddle steamer. She was a very sturdy sea boat built by Denny's of Dumbarton and no doubt would have fared well in preservation. Instead, she was to become a floating pub and restaurant in London. (Douglas McGowan collection)

Waverley departs from Rothesay for an afternoon cruise to Tighnabruaich, August 1966. Note the lions adorning the funnels. This was her second season with blue hull and chocolate boot-topping. (John Goss)

Departing Brodick for an afternoon cruise to Pladda with Arran's highest peak Goatfell in the background, 1 September, 1969. Over the years, Brodick has consistently proved to be a lucrative pick-up point for *Waverley* on her afternoon cruises. (John Goss)

Alongside Tighnabruaich pier, September 1969. This has been one of *Waverley*'s most popular destinations throughout her career and still features prominently in her timetable today. (Douglas McGowan)

Although the Clyde steamer fleet was historically laid up for the winter in Greenock harbours, this practice was changed in the late 1960s. *Waverley* is berthed alongside *Queen Mary II* with *Caledonia* alongside *Duchess of Hamilton* in Glasgow's Queens Dock, November 1967. (Douglas McGowan)

At Rothesay pier at low tide, raising steam ready for departure to Dunoon, Gourock and Craigendoran, August 1970. The blue hull is gone and *Waverley* is again sporting a black hull, but retaining the lions on the funnels. (Douglas McGowan)

On the twenty-fifth anniversary cruise, Terry Sylvester's two-year-old daughter Sharon blows out the candles on the special cake, watched by John Whittle, general manager of the Caledonian Steam Packet Co. (Bill Dempster)

Right: Captain McEachran of the *Waverley* has the honour of plunging the knife into the *Waverley*'s paddlebox on 19 May 1972. The cake was specially made by the CSP Co.'s own Gourock bakery which provided all the scones, pancakes, sausage rolls and cakes for the CSP fleet. (Bill Dempster)

Opposite: Jean Martin (to become Mrs McGowan!) unveils a brass plaque from the Paddle Steamer Preservation Society to commemorate *Waverley*'s twenty-fifth anniversary on a special cruise on 19 May 1972. The historic railway silverware on the table was in turn donated to the PSPS by the Caledonian Steam Packet Co. (Bill Dempster)

Waverley approaches Wemyss bay, the PSPS flag flying proudly from the mainmast, 29 September 1973. This was a special charter sailing by the PSPS around the lochs and, although not known at the time, was to be the steamer's penultimate day in service for Caledonian MacBrayne Ltd (CalMac). (John Goss)

Opposite top: *Waverley* passing Ardbeg on her final approach to Rothesay on the return leg of the PSPS charter, 29 September 1973. (John Goss)

Opposite below: Approaching Millport during her last days of service for CalMac, September 1973. In this year, her funnels were painted red with black tops, retaining the CSP lions. This was to be the new colour scheme of Caledonian MacBrayne, an amalgam of David MacBrayne Ltd and the Caledonian Steam Packet Co. Ltd. (John Goss)

Waverley's purser, John Brewster, being presented with an engraved cigarette lighter by the Scottish branch of the PSPS on 29 September 1973. The presentation is being made by Jean McGowan. (John Goss)

Final days under the CalMac flag: Douglas McGowan persuaded Captain Munro, officers and crew to pose for the camera at Millport pier whilst *Waverley* was on PSPS charter, 29 September 1973. (John Goss)

Waverley arriving at Dunoon from Rothesay in her final few weeks of CalMac service, September 1973. (Douglas McGowan)

Waverley's stokehold, September 1973. Originally coal-fired when built, she was converted to oil burning in the mid-1950s. Re-boilered in 1981 and again in 2000, *Waverley* now has twin boilers, which allows for greater operational flexibility. (John Goss)

The CalMac officers and crew in the early 1970s: Captain Hugh Campbell keeps a watchful eye on the Bridge. (Douglas McGowan)

John Ellis, chief officer. John had been 'mate' for many years on PS *Caledonia*. Note the Bridge telegraphs painted black to avoid polishing! This was one of the first tasks which the PSPS work parties addressed some months later. (Douglas McGowan)

Pat McGhee, donkeyman/greaser, oil can at the ready! (Douglas McGowan)

Dan and 'China' in the stokehold, 1973. 'China' gets a mention in Billy Connolly's biography by Pamela Stephenson. (Douglas McGowan)

One of the Caledonian Steam Packet Company's old stalwarts: Roddy on the capstan, pictured at Millport, August 1972. Roddy had previously worked aboard *Caledonia* for several years prior to her withdrawal in 1969. (Douglas McGowan)

Chief Steward Angus Kennedy in the ship's dining saloon. Right up until her final days with CalMac, *Waverley* offered traditional waitress service for lunch and high tea which was booked in advance at the Purser's Office. Normally, there were three sittings for both lunch and high tea. (Douglas McGowan)

Purser John Brewster, who gave the PSPS much support in their campaign to preserve the steamer. (Douglas McGowan)

'The galley boys'! *Waverley*'s galley, August 1973. Note the original coal-fired stove in the background, unchanged since 1947. This was later inefficient and was difficult to maintain, although WSN continued to use it in the early years of preservation before converting to Calor gas. (Douglas McGowan)

The tugs position themselves for *Waverley*'s short tow upriver from the James Watt Dock, Greenock, to Lamont's slipway at Port Glasgow on 21 February 1974. (James Hall Photographers)

Waverley is towed stern first from the James Watt Dock into the main channel, 21 February 1974. The survey she would undergo on the slipway would determine her future. (James Hall Photographers)

A friendly good-luck wave from two locals as *Waverley* passes Port Glasgow and approaches the slipway of James Lamont. Captain Hugh Campbell was aboard, representing Caledonian MacBrayne Ltd, and Douglas McGowan and Terry Sylvester for the PSPS. (James Hall Photographers)

Waverley arrives at the slipway, PSPS flag flying from the mainmast (although ownership was still with CalMac), 21 February 1974. The tugs, having done their job, are getting ready to cast off and already a line from the port bow is secure on the berthing jetty. One of Lamont's men is in a small boat ensuring the vessel is properly lined up. Sir William Gray, Lord Provost of Glasgow, poses for television and press cameras as he 'pulls' *Waverley* up the slipway. (Norman Burniston Photography)

Douglas McGowan no doubt pondering the future as *Waverley* slowly rises out of the water, pulled by giant steel hawsers on a pulley system (under *Waverley*'s bow). The process from start to finish took just over an hour. On 21 February 1974. (Norman Burniston Photography)

A proud moment! John Whittle, deputy chairman and chief executive of Caledonian MacBrayne Ltd, signs the ship's papers and legal documents in CalMac's Gourock office, 8 August 1974. Looking on are Colonel Sir Patrick Thomas, chairman of the Scottish Transport Group, Douglas McGowan and Terry Sylvester (WSN). (James Hall Photographers)

The ship's papers are duly passed to a somewhat bemused Douglas McGowan who was to become the new Ship's Husband, 8 August 1974. (James Hall Photographers)

With all the formalities duly completed and papers signed by both parties, Douglas McGowan hands over that famous pound to Pat Thomas. History is being made! (Note: What did the brown paper parcel on the right of the photograph contain – was it Terry Sylvester's sandwiches?!)

The party, pursued by press and television crews, then went outside on to Gourock pier to re-enact the ceremony once again … and yet again at the James Watt Dock where the steamer was berthed! (George Young Photographers)

Scottish Television personality Bill Tennant launches the public appeal at a press conference in *Waverley*'s dining saloon, 25 October 1974. From left to right: George Train, Douglas McGowan, Bill Tennant, Terry Sylvester and Peter Reid. (James Hall Photographers)

Bill Tennant donates a personal cheque to 'kick-start' the public appeal which within a few months had raised £100,000. However, even at this time, the steamer's fanboards proudly proclaiming an afternoon cruise to Gourock, Dunoon, Wemyss Bay, Largs, Rothesay and the Kyles of Bute seemed but a distant dream. 25 October 1974. (James Hall Photographers)

An open day was held in November 1974 to maintain public awareness. This aroused considerable interest from both public and media and to mark the occasion, the ship was dressed with flags, although the funnels were still covered. It would be another six months before smoke would emerge from the giant twin funnels once again. (James Hall Photographers)

Douglas McGowan and Terry Sylvester keep a watchful eye from the ship's bow as *Waverley* is gently coaxed into the Scott's Garvel dry dock at Greenock for a major survey and refit, 17 February 1975.

The final stages of *Waverley*'s refit and overhaul. The vessel has been refloated and returned to the James Watt dock where real gold leaf is being applied to the scroll work on the paddlebox surrounding Edward Waverley, 2 May 1975. (James Hall Photographers)

A Scottish television film crew interview Douglas McGowan on the dockside at Greenock, as the steamer has the finishing touches applied in the background, May 1975. (James Hall Photographers)

Yet another interview! This time, in *Waverley*'s wheelhouse being interviewed by reporter Les Wilson of STV with Bill Summers on the left of the photograph, 3 May 1975. (Norman Burniston Photography)

Yet another proud moment! Having left the James Watt dock far behind, *Waverley* is moving again for the first time in nineteen months. She is off Kilcreggan, undergoing trials and compass adjustments, with the brand new Waverley Steam Navigation Co. houseflag, based on the old LNER houseflag and St Andrew's cross, flying from the mainmast for the first time. May 1975. (James Hall Photographers)

Waverley gets ready to receive her first passengers for the inaugural cruise on Thursday 22 May 1975. The flags are flying but the bulldozer has only just finished tidying up the quay before the 2.00 p.m. departure with VIP guests. The Waverley Terminal sign has only been in position for an hour, courtesy of Glasgow District Council! (John Goss)

Waverley settles into a regular routine from her new base at Glasgow's Anderston Quay, May 1975. (Douglas McGowan collection)

Bill Summers, *Waverley*'s chief engineer from 1947 until 1969, returns to his old ship in February 1975 to assist in the engine overhaul at the Garvel dry dock. Bill's wealth of experience was invaluable, especially in the first few months of preservation. (Douglas McGowan collection)

GREENOCK TELEGRAPH, Friday, 9 August, 1974

SIR Patrick Thomson accepts a pound for the Waverley from Mr Douglas McGowan while a smiling Mr J. Whittle looks on. On the extreme right is another member of the Paddle Steamer Preservation Society.

Veteran paddle steamer changes hands for £1

VETERAN paddle-steamer Waverley changed hands yesterday for only £1.

But it will cost her new owners, the Paddle Steamer Preservation Society, £100,000 to get her shipshape if they intend to put her back into service on the Clyde.

During a light-hearted ceremony at Caiedonien MacBrayne headquarters at Gourock, Sir Patrick Thomas, chairman of the Scottish Transport Group, gave the Society the pound with which to pay for the Waverley.

To make transfer of ownership legal it was necessary for the ship to change hands for a nominal sum. This means that she is not an out-and-out gift and that the steamer company can impose certain conditions regarding her future.

Mr John Whittle, general manager of Caledonian MacBrayne, handed over the ship's documents to Mr Douglas McGowan, honorary secretary of the society.

FINE GESTURE

Mr McGowan, the old steamer's champion and one of those striving to see she will sail again, said it was a magnificent gesture.

"As a child I always admired the Waverley — little did I know that one day I would own it with some friends," he laughed.

"The Clyde badly needs a special tourist attraction. Having this paddle steamer is a priceless asset.

question of the Waverley's future

had occurred in any other country, the steamer would already have been put to its full potential as a tourist attraction.

"If the society gets its way, the Waverley will sail on the Clyde next summer," he promised.

He added that it would be necessary to await a decision of the Strathclyde Regional Council

over whether they will subsidise

Left: *Greenock Telegraph*
Below: *Glasgow Herald*
Bottom: *Daily Mail*

The paddle steamer Waverley, which may sail again on the Clyde.

Waverley may sail again

BY A STAFF REPORTER

The world's last seagoing paddle steamer, the Waverley, which was taken out of service last year, may sail again on the Clyde under new colours.

Her fate depends on an

cost to make her seaworthy.

The owners, Caledonian MacBrayne, have offered the Waverley to the Paddle Steamer

spending about £1500 to assess her future.

The steamer will be moved tomorrow from the James Watt Dock, Greenock, to Port Glasgow, where she will be

If, however, the hull will keep her afloat safely without extensive maintenance for some years, a public appeal may be launched for funds to keep the Waverley as a special

Daily Mail, Friday, August 9, 1974 — 8 — PAGE

Paddling on ... the steamer with a £1 price tag

The Waverley—cruising down the Clyde.

THE last of Scotland's famous seagoing paddle steamers changed hands yesterday . . . for £1.

And now, instead of ending her days in the breakers' yard, the old lady of the Clyde will live on in dignity — and may even become a tourist attraction.

The Waverley, built in 1947 at a cost of £145,000, was taken over by the Paddle Steamer Preservation Society from her previous owners Caledonian MacBrayne.

But the £1 price tag is only a token payment to legalise the transfer. Her launching as a holiday cruiser will cost an estimated £100,000.

Financial help is being sought by the society and already the wheels, if not the paddles, are turning.

Cash has been pledged by various tourist boards in Scotland, and the Highlands and Islands Development Board have declared an interest in keeping the vessel afloat.

But the best chance of 'full ahead for'ard' for the Waverley may be a move by Sir William Gray, Lord Provost of Glasgow.

He has asked the new Strathclyde Regional Council to set up a sub-committee to investigate the possibility of including the boat's running costs in their annual budget.

So the Waverley, which last sailed the Clyde commercially in 1973, may soon be casting off again . . . this time on the rates.

Battle to save the Waverley begins

FIRST SHOTS in the battle to save the paddle steamer Waverley, now laid up at the James Watt Dock in Greenock, have been fired by the Scottish branch of the Paddle Steamer Preservation Society.

Secretary of the branch, Mr Terry Sylvester, points out in a letter to the *Telegraph* that the 26-year-old steamer — the last paddle steamer to be built for passenger services in Britain and the last salt water paddle steamer to ply in Europe — had world-wide publicity last year.

He points out that the Waverley's stardom started when the ship was depicted on the front-cover of a publication produced by the Clyde tourist authorities.

Says Mr Sylvester: "She followed by being the subject of an article in the *Sunday Times* — published nationwide; was filmed by BBC Television for a programme shown throughout Britain in November, and was used for film-making by a National Film Theatre group.

"BBC radio recorded her famous band for a national programme and Scottish Radio also featured her, making recordings on board.

FILMED

of the season and the Paddle Steamer Preservati[...] had 500 posters fe[...] ship distributed al[...] British Isles."

Mr Sylvester adds th[...] Tourist Authority are v[...] in publicising the Waver[...] both at home and overs[...]

"All this publicity," h[...] be of tremendous ben[...] tourism. She present[...] spectacle at all the Clyd[...] creates tremendous int[...] the watching tourists.

"Although the Clyde[...] other ships sailing its w[...] of these has created[...] publicity.

LAST OF TY[...]

"The Waverley succ[...] she is unique — last of h[...] world. No other tourist[...] claim to a real worki[...] paddle steamer.

"This letter is principa[...] everyone concerned w[...] perity of the West of Scot[...] tourism to make cap[...] 'working dinosaur' at[...] tunity for 1974.

"The Waverley will[...]

Greenock Telegraph

Waverley given to Preservation Society

Gift is a challenge to us, says secretary

By a "Telegraph" Reporter

WHAT do you do when you are suddenly handed a 693-ton "baby" and told—"It's all yours"? That is the position that the Paddle Steamer Preservation Society finds itself in today.

Only, in this case, the "baby" is the 26-year-old paddle steamer Waverley, which Caledonian MacBrayne have handed them as a pre-Christmas present.

It is an expensive gift, too, as the ship could have fetched £15,000 even as scrap — and it is known that an English brewery firm was interested in buying her to turn her into a pub, just like her [...]

working closely with the Maritime Trust in London," said Douglas.

The gift was the steamer company's way of saying "Thank You" to the society for all the publicity it had geared up to attract tourists to the Clyde — and the Waverley.

One thought remains. Where will the members get their drive from now that the last sea-going paddler has gone?

"We've still got the Maid of the Loch," laughed Douglas.

Shivering Scots face cuts tonight

AS SNOW-STRUCK Scotland shivered today, more than 300,000 consumers were warned that they faced electricity cuts during the evening peak period.

There would also be voltage reductions. The Board warned that these could come early today — possibly at lunchtime.

The North of Scotland Hydro-Electric Board also warned of possible voltage reductions this afternoon.

In Scotland, all minor roads in the west were ice-bound this morning while snow was still falling in the east. In Perth and Angus, roads were passable only with care.

This is how the Paddle Steamer Preservation Society would like to see their new acquisition, the Waverley—sailing down the Clyde with their flag flying at the masthead. The picture was taken during one of their cruises this summer.

The ex-Clyde steamer Waverley, thought to be the world's last sea-going paddle steamer, lies high and dry yesterday on the slipway at Lamont's shipyard, Port Glasgow, where she will undergo maintenance work and a Department of Trade and Industry survey. If the department's report is favourable the Paddle Steamer Preservation Society and Clyde Tourist Association hope to return her to service on the Clyde as a summer attraction

● Tugs manoeuvre the paddle steamer Waverley from the James Watt Dock today at Greenock on the journey to Lamont's yard.

Bid to keep Waverley afloat

By John McCormick

A BID to keep the world's last sea-going paddle-steamer, sailing in Scottish waters has been made by Rothesay Harbour Trust.

The 27-year-old paddler Waverley, which used to sail the Clyde loaded with trippers, has been taken out of service by her owners, Caledonian MacBrayne Ltd., but there are no plans to scrap her.

At a recent meeting of the trust, the offer of a berth for the Waverley at Rothesay was made with the suggestion that the old paddler could still sail from there on mini-cruises.

Last night Rothesay's provost, Mr. Donald McPhail, said: "It would be wonderful to have the world's last sea-going paddler still sailing in Clyde waters.

"If she was sold to the Paddle Steamer Preservation Society she could still sail. It would not be in opposition to Caledonian MacBraynes."

Mr. Douglas McGowan, of the preservation society, commented later: "We are having talks at the moment with the Clyde Tourist Association, the Highlands and Islands Development Board and the Scottish Tourist Board, with a view to arranging subsidies for the Waverley's operating losses.

Will the Waverley sail again?

LORD PROVOST William Gray lent a willing hand at a capstan today to help winch the veteran Clyde paddle steamer Waverley on to the slipway at Lamont's shipyard, Port Glasgow, for a £3000 survey.

The Waverley, the world's last seagoing paddler, has been offered by Caledonian MacBrayne to the Paddle Steamer Preservation Society for use as a cruise ship.

But her fate—whether she will cruise again or become a static museum — depends on the survey, the result of which should be known to the society in about 10 days.

The Clyde Tourist Association and other groups are considering financial assistance for the project.

Lord Provost Gray, who is chairman of the association, was at the yard today as the Waverley was towed by two tugs on the 45-minute journey from the James Watt Dock at Greenock.

He said—"We want to get a firm report on the actual condition of the ship by independent surveyors so that we kn[illegible] exactly the cost involved in [illegible]

"As Lord Provost, like so many Glaswegians who have sailed in paddle steamers over the years, I think it would be sad if the time came when we did not have a single paddle steamer on the Clyde.

"It is part of the heritage not just of the West of Scotland but of Glasgow."

● Mr Douglas McGowan (left), secretary of the Paddle Steamer Preservation Society, aboard the Waverley today with skipper Mr Hugh Campbell.

Above: Scottish *Daily Express*
Left: *Evening Times*

Opposite Page, clockwise from top left: Greenock Telegraph; Daily Record; Scotsman.

LORD PROVOST GRAY (centre) helping to winch the Waverley in.

Lord Provost puts his weight behind the Waverley!

Appeal launched to save the Waverley

TELEVISION personality Bill Tennent yesterday handed the Waverley Steam Navigation Company Ltd. a cheque for £10 and in so doing, launched an ambitious public appeal which, it is hoped, will raise enough funds to allow the veteran paddler to resume her career as a sea-going pleasure steamer.

Whether the Waverley will again convey holidaymakers on summer excursions to the Clyde resorts depends entirely on response to the appeal which is aiming at a target of £50,000.

If the appeal flops, the steamer could end her days as a maritime museum.

BUNTING

The Waverley's bunting fluttered brightly in the breeze at James Watt Dock yesterday as Mr Douglas McGowan, a director of the ship's owners, spelled out their plans for the steamer's future.

Speaking on board the ship he said: "Glasgow Corporation have assisted in the meantime with a grant of £6,000, but to put the ship back into service, we need about £50,000. Our own funds now total over £20,000 and the Scottish Tourist Board and the Highlands and Islands Development Board have pledged their financial support.

"We are meeting Strathclyde's Leisure and Recreation Committee on board this afternoon and we look to them for assistance in operating the ship."

Mr McGowan said their plan was to put the steamer into service between May and September. She would undertake special charters during May, June and September and public sailings during July and August.

The Paddle Steamer Preservation Society would remain owners and she would be managed on their behalf by Caledonian MacBrayne or another operator who would supply a crew.

REPAIRS

Waverley sea-going again. But after hydraulic tests had been completed this week, the boiler was found to be in good repair and this had reduced the total funds necessary.

Today is "open day" on board the ship when the public have the chance of seeing for themselves how work is progressing. One striking difference is the Waverley's funnels which have been repainted in her former LNER colours — red and black with a white band.

"We believe that the general public would like to see the Waverley sail again and there can be no doubt that she would be a tremendous boost to tourism. If our plan works the Firth of Clyde will have a really unique attraction — something really different that will generate its own market as time goes on," said Mr McGowan.

Provost in bid to save paddle boat

A LAST ditch bid is to be made by Lord Provost William Gray of Glasgow to save the paddle steamer Waverley.

He is going to ask the Government for a direct subsidy to keep her sailing on the Clyde because of her importance to tourism.

But it will cost £50,000 just to carry out repairs—and after that an increasing burden of rising costs would have to be met.

The Lord Provost, chairman of the Clyde Tourist Association, said they had already received an offer of £20,000 towards the initial cost.

He said: "The Isle of Man runs old steam loc... as a tourist attraction.

"Keeping a paddle steamer on the Clyde would be an enormous and valuable attraction for us."

After the association met in Gourock yesterday Lord Provost Gray said it would be comparatively easy to raise the extra £30,000 by public appeal.

Subsidy

Move to save Waverley

By STAFF REPORTERS

The Clyde Tourist Association are to investigate the possibility of buying the Clyde paddle steamer Waverley, which was put up for sale yesterday by the board of the Scottish Transport Group.

At their annual general meeting in Glasgow yesterday, the finance of the local authorities in the Strathclyde region to keep the steamer going for a year once the reorganisation of local government came into effect in May.

However, the members of the tourist association said they were against taking a

WAVERLEY APPEAL

TOTAL DONATIONS NOW STAND AT £21,000

PLEASE HELP TO PRESERVE THE LAST SEA-GOING PADDLE STEAMER IN THE WORLD — ON THE CLYDE.

SEND YOUR DONATION NOW TO:—

WAVERLEY APPEAL, P.O. BOX, GLASGOW

I am proud to contribute £.......... towards the Paddle Steamer Preservation Society's appeal for the preservation of the Paddle Steamer Waverley. I understand that if preservation fails the unexpended part of my contribution will be refunded on request.

Signed.............................Date..................1974

Address (block capitals please)..............................

Mr William Gray, Lord Provost of Glasgow, helping to turn the capstan to winch the veteran Clyde steamer Waverley on to the slipway yesterday at Lamont's shipyard, Port Glasgow, where she is to undergo a £3000 survey. (Report on Back Page.)

Terry Sylvester (WSN chairman) welcomes on board the VIP guests on 22 May 1975. Looking on are (from left to right): Iain Dunderdale (chief officer), Cameron Marshall (assistant purser), Douglas McGowan and Peter Reid (WSN directors). (John Goss)

Sir William Gray, Glasgow's Lord Provost, replies on behalf of the guests, as the steamer gets ready to cast off. (John Goss)

A beaming Terry Sylvester accepts a cheque for £12,500 from Philippe Taylor, chief executive of the Scottish Tourist Board. A further cheque for the same amount followed. 22 May 1975. (Scottish Daily News)

The Strathclyde Police Pipe Band give *Waverley* a good send-off as she casts off on her 'second maiden voyage' from Glasgow's Anderston Quay. 22 May 1975. (John Goss)

White Horse Whisky Bar

Waverley's first full public cruise was a special charter by the PSPS on Saturday 24 May from Glasgow to Gourock, Tarbert and Ardrishaig. She is seen here with a good complement of passengers, arriving at Gourock. Special permission was obtained from CalMac to use their pier on this occasion with Greenock Customhouse Quay becoming *Waverley*'s regular port of call in later years. (John Goss)

Opposite top: Having already disembarked most of her guests, *Waverley* leaves Dunoon pier on a photographic short cruise along the Cowal coast during her first ever excursion under the WSN flag. 22 May 1975. (John Goss)

Opposite below: The newly refurbished White Horse Whisky Bar in May 1975. White Horse Distillers were the first company to offer to help restore the ship and went on to sponsor this bar for several seasons thereafter. Prior to the refurbishment, this area of the ship was a rather dull and dingy corner, lacking any character. Today, it is one of the most popular and cosy lounges on the ship. (Douglas McGowan collection)

The crowds get ready to surge aboard as *Waverley* berths at Gourock pier on 24 May 1975. (John Goss)

The paddle steamer Waverley leaving Anderston Quay yesterday on a special cruise to Dunoon.

Clyde's cruising potential

By CLAUDE THOMSON,

As the veteran paddler Waverley prepared to start her first sail down the Clyde in her new life as a pleasure steamer, Strathclyde Regional Council's leisure and recreation committee yesterday decided to investigate the potential of cruising on the Clyde and West Coast.

A report covering not only the Clyde estuary and sea lochs but Loch Lomond and Loch Awe, is to be prepared by Mr Arthur Oldham, director of leisure and recreation, and Mr Ian McFarlane, director of policy planning.

The Highlands and Islands Development Board and the Scottish Tourist Board will co-operate with the region in compiling the report.

A cautionary note against unduly high expectations of financial

Left: *Glasgow Herald*

Below: *Daily Record*

STEAMING BACK ON CREST OF A WAVE

Story: ROY TOWERS   Picture: JAMES ROBERTSON

A CLYDE tradition is kept alive as the grand old paddle steamer Waverley thrashes her way from Glasgow to Dunoon.

With flags flying and 400 guests crowding her sunlit decks, the 28-year-old steamer slipped away yesterday from Anderston Quay on the first of a summer series of cruises "doon the watter."

BOOST

And the first thrash of the giant paddles was as good as a 10-gun salute for the band of dedicated paddle steamer enthusiasts who fought to keep the Waverley working.

Last year the old ship seemed destined for the scrap yard when Caledonian-MacBrayne put her up for sale.

But eventually the conservationists bought her for a nominal £1, and set about keeping her seaworthy.

That set them back £80,000.

Now they're called the Waverley Steam Navigation Company and just before yesterday's cruise they got a £12,500 boost from the Scottish Tourist Board.

The money will go to repairing and renovating the paddle steamer. Strathclyde Regional Council have also promised £30,000.

The Waverley has the distinction of being the last sea-going paddle steamer in the world.

She was built in 1947 and after working for the L N E R railway company and British Rail she was moved to Caledonian-MacBrayne.

Every weekend in July and August she will make sailings from Anderston Quay down the river.

And every weekday except Friday, the trips will be from Ayr, up the Clyde.

Sample trips from the

Waverley's timetable include a day trip on Sunday, May 25 from Glasgow to Greenock, Dunoon, Largs, Rothesay or Millport with return adult fares ranging from £1 for Greenock up to £2.25 for Largs and Millport.

Children under 16 go for 50p to Greenock and £1 on the longer trip.

RETURN

On Monday, a public holiday, a one-off trip is planned around Arran to Campbeltown with an adult return fare of £3.95.

On Wednesday, May 28, a special cheap day excursion will take Glasgow trippers round the Holy Loch with an hour at Dunoon for £1.95.

Arriving at Ardrishaig, *Waverley* lists to port as the 700 passengers await disembarkation. 24 May 1975. (John Goss)

Waverley makes a fine sight as she basks in the sunshine at Ardrishaig where she embarked a number of passengers for an afternoon cruise on Loch Fyne. 24 May 1975. (John Goss)

Departing from Ardrishaig for the return trip to Tarbert, Gourock and Glasgow on 24 May 1975. There was no indication at this stage of the calamity which was to follow. Only half an hour later, she broke a paddle float, resulting in a major delay to the schedule. She eventually arrived back in Glasgow at 3.55 a.m. next morning. (John Goss)

William Tennant's Old Tyme Steamer Band! This band had been entertaining passengers on the Clyde steamers for years and was keen to revive the tradition on *Waverley* in her first season. Pictured at Gourock pier, May 1975. (Douglas McGowan collection)

Above: The last sea-going paddle steamer in the world proudly leaves Glasgow's Anderston Quay on one of her early excursions as a preserved ship. June 1975. (Douglas McGowan collection)

Left: Jean and Douglas McGowan aboard *Waverley* with their sixteen-month-old daughter Lynn. June 1976. (Second City Films)

Opposite below: The Glasgow music-hall entertainer Glen Daly supported the ship by recording a song called *Save, Save the Waverley*. Whilst not quite reaching number one in the charts, it did sell well in the *Waverley*'s own souvenir shop! May 1976. (Second City Films)

Leaving Ayr Harbour for the first time, June 1975.

Waverley passes through the narrows of the Kyles of Bute in June 1975. This photograph shows clearly that there is little margin for error in this narrow channel. (Ian Shannon)

Waverley and *Queen Mary II* pass near Colintraive in the Kyles of Bute. June 1975. (Ian Shannon)

Opposite below: Berthing at Dunoon on her return from cruising to Tighnabruaich, June 1975. (Ian Shannon)

Waverley's first call at Millport in WSN livery with the MV *Keppel* arriving from Largs. 26 May 1975. (Ian Shannon)

The steamer has now backed across to the south wall and passengers disembark. In 1975, Ayr was still a bustling fishing port as can be seen by the numerous fishing boats. Sadly, the scene today is quite different and what remains of the fleet has been transferred to Troon. The Mission to Deep Sea Fishermen building has been demolished and the area is now occupied by a block of flats.

Opposite above: In her first season, *Waverley* carried out a number of special excursions, one of which was from Ayr to Girvan and round Ailsa Craig. This cruise was a great success and sold out completely, as can be seen from the large crowd on the harbour wall. June 1975. (Ian Shannon)

Opposite below: *Waverley* canting in the north basin at Ayr harbour. This is a difficult manoeuvre, even in perfect conditions for a paddle steamer but when there are strong winds and fresh water coming down the river it really tests the navigational skills of *Waverley*'s Master. The ship is about to be pulled round to port by the use of the cant rope so that her stern is facing up the river. The small boat seen at *Waverley*'s port bow was used in the early weeks of 1975 to assist with ropes and, if necessary, 'nudging' the bow to assist in lining her up properly prior to berthing. June 1975. (John Goss)

Waverley's first season in preservation certainly uncovered a number of 'characters', none more so than Alec McLeod from the Isle of Lewis, deckhand extraordinaire, seen here in fancy dress at Campbeltown. (Douglas McGowan)

Waverley arrives at Tighnabruaich pier, with a good complement of passengers, June 1975. (Ian Shannon)

Approaching Stranraer, in Loch Ryan. This was an unusual port of call for *Waverley* and the general public and enthusiasts alike supported the cruise in big numbers. 29 June 1975. (John Goss)

A pensive Douglas McGowan surveys the scene at Kilcreggan on a summer Saturday in 1975. (Second City Films)

Waverley manoeuvring at the first Millport call under the WSN flag on Monday 26 May 1975. The cruise was to be curtailed at Millport due to paddlewheel problems. Millport is never the easiest of piers for a paddle steamer, the approach being dominated by rocks. At low water, the pier itself has barely sufficient depth of water to accommodate *Waverley*. (John Goss)

Waverley's paddlebox, May 1975. As the paddlebox – with the paddles thrashing the white foaming water approaching or departing a pier – is the part of the ship everyone loves to watch, there was a considerable focus on this area of the ship in an effort to make it look as attractive as possible. (John Goss)

A rather forlorn *Waverley* is towed the short distance from Anderston Quay to Govan dry dock, February 1976. The *Sunday Mail* spearheaded the '10 days to save the *Waverley*' campaign, which enabled the company to stay afloat for another year. (*The Herald* and *Evening Times* © SMG Newspapers)

Waverley departs Rothesay for the Kyles of Bute with MV *Glen Sannox* approaching the pier from Wemyss Bay. Note: *Waverley*'s port aft wooden lifeboat has gone, replaced by inflatable liferafts to increase passenger capacity, August 1976. (Ian Shannon)

Waverley departs Gourock pier, June 1976. Today, Gourock pier is a mere shadow of its former self and not much more than the Dunoon car ferry berth remains. (Ian Shannon)

Thundering through the Narrows of the Kyles of Bute, returning to Rothesay from Tighnabruaich. The flags flying above the wheelhouse are 'The Sunday Mail' and 'White Horse Whisky', both of whom remained two major sponsors in 1976. (Ian Shannon)

As dusk begins to fall, *Waverley* sits rather ignominiously at the unfamiliar surroundings of Dunglass oil terminal, Old Kilpatrick. She has developed a major paddlewheel failure and passengers have been transferred to TS *Queen Mary* which berthed alongside *Waverley* for the upriver cruise back to Glasgow, 7 August 1976. (Ian Shannon)

With the remains of the old Govan pier on the left, *Waverley* makes her way downriver. It is hard to believe that *Waverley* is now the only ship to regularly use the upper reaches of the Clyde. In this photograph, a number of ships are in evidence and MV *Glen Sannox* can be seen making her way upriver for a special charter, July 1976. (Ian Shannon)

On a beautiful August day in 1976, *Waverley* makes an impressive sight as she sweeps round the head of Loch Riddon with the disused pier of Ormidale in the background. (Ian Shannon)

1. *Caledonia* arriving back at Gourock from a cruise to Arran and Pladda, August 1968. (Douglas McGowan)

2. *Waverley* sweeps into Rothesay bay at full speed, August 1972. (Douglas McGowan collection)

3. *Waverley* berthing at Rothesay, June 1971. (Joe McKendrick)

4. A seagull uses *Waverley*'s 'stump' foremast as a convenient perch as she approaches Dunoon with a good load of passengers, August 1971. (Lawrence Macduff)

5. *Waverley*'s appearance for the 1973 season in her new CalMac livery was quite pleasing to the eye. Note the Caledonian Steam Packet pennant still being flown from her mainmast, although she was now owned by Caledonian MacBrayne Ltd. May 1973. (Lawrence Macduff)

6. This colour scheme, pictured at Lamont's of Port Glasgow, could certainly not be described as pleasing to the eye. It was changed the following day to red with black top. March 1973. (Douglas McGowan)

7. *Above: Waverley*'s dining saloon in the final year of CalMac service, 15 September 1973. (Lawrence Macduff)

8. *Right:* The same dining saloon twenty months later, ready for *Waverley*'s inaugural cruise for WSN. Jean McGowan was hard at work making the curtains until 3.00 a.m. the previous morning! 22 May 1975. (Douglas McGowan)

9. *Waverley* begins to undergo transformation following the change of ownership. At the James Watt Dock, Greenock, her funnels are painted in the old LNER colours of red, white and black. September 1974. (Douglas McGowan)

10. PSPS work parties proudly pose for the camera on 17 August 1974. The steamer had only changed hands a few days previously – it didn't take long for members to get their hands dirty. (John Beveridge)

11. *Waverley* berths at Ardrishaig on her first public cruise for WSN, 24 May 1975. (Lawrence Macduff)

12. Leaving Tarbert, Loch Fyne, 24 May 1975. (Lawrence Macduff)

13. Third day in service as *Waverley* arrives at Dunoon on 25 May 1975. (Lawrence Macduff)

14. A packed *Waverley* makes her way serenely up the Kyles of Bute on 6 July 1975. The flags of White Horse whisky and Esso petroleum (a deadly concoction!) are being flown from the foremast. (Lawrence Macduff)

15. Approaching Millport with an impressive bow wave and another good complement of passengers, 7 July 1976. (Lawrence Macduff)

16. *Waverley* berthed at the unusual location of the coal pier, Dunoon, on the day following her grounding on the Gantock rocks. (James Aikman Smith)

17. Another view of the stricken vessel at the coal pier. The starboard (aft) lifeboat is in the process of being repositioned on the davits. (James Aikman Smith)

18. Just proving she really is a steamship! Some old socks must be burning in the stokehold as *Waverley* departs from Glasgow, back in service following the Gantocks grounding. 3 September 1977. (Lawrence Macduff)

19. *Waverley* at her Glasgow berth, 7 September 1977. MV *Queen of Scots*, which had played such a vital role in maintaining some continuity during the preceding weeks, lies ahead of *Waverley*. (Lawrence Macduff)

20. Berthing at Irvine on a special excursion from Ayr, June 1978. (James Aikman Smith)

21. Arriving back at Brodick from Campbeltown, August 1994. (Douglas McGowan)

22. If sea conditions allow, *Waverley* occasionally has a scheduled call at Lynmouth, en route to Ilfracombe. The anchor is dropped and passengers are landed by a small launch, allowing them the opportunity to explore this delightful Devon village. (Douglas McGowan)

23. *Waverley* at Dunoon on a beautiful Clyde evening, 20 July 1996. (Joe McKendrick)

24. *Waverley* anchored at Lundy Island in the Bristol Channel, June 1997. A small pier has now been constructed which allows much faster passenger disembarkation and more time ashore. (Douglas McGowan)

25. Arriving at Rothesay, Isle of Bute, July 1997. (Gordon Wilson)

26. *Waverley* makes a majestic arrival at Largs, August 1999. (Douglas McGowan)

27. It's 7.00 a.m. in Oban Bay on 5 May 2002 and the tranquility of an idyllic early Sunday morning in spring is briefly disturbed by the gentle paddlebeats of *Waverley* as she heads for Fort William. (Douglas McGowan)

28. A couple of minutes later and *Waverley* leaves the Island of Kerrera behind and heads towards Loch Linnhe for the morning cruise to Fort William. (Douglas McGowan)

29. *Waverley* arrives at Largs in perfect weather for the cruise to Tarbert, Loch Fyne, on 13 August 2002. (Douglas McGowan)

30. Full astern! Departing from Tarbert for an afternoon cruise on Loch Fyne, later on the same day. (Douglas McGowan)

31. *Waverley* proudly arrives back on the Clyde from Great Yarmouth following stage one of her rebuild, 18 August 2000. (George Young Photographers)

32. The next incarnation. *Waverley* sits high and dry on a floating pontoon at Great Yarmouth, minus masts and forward deck shelter, as stage two of her major rebuild project gathers pace. In total, the ship will have had over £7 million spent on her since 2000 to improve her passenger and crew amenities and allow her to sail well into the twenty-first century. (Douglas McGowan)

33. *Waverley's* funnels, with Walter Bowie, the ship's Junior Purser, reflected in the window to the left. (Alistair Deayton)

Above: General Manager Bob Wilson (third from left) in conference with Chief Engineer Bill Miller (left). Chief Officer Neil McLeod is on the right. (Second City Films)

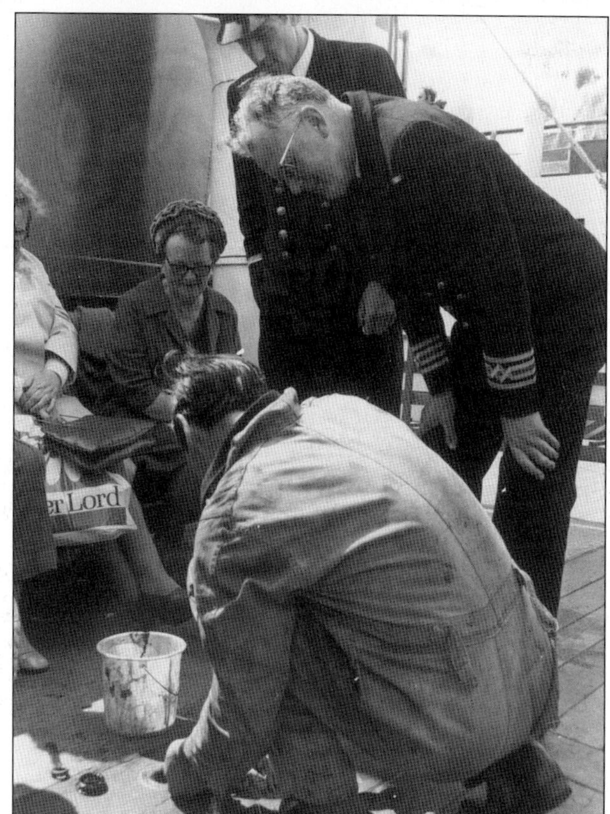

Right: 'Dipping the tanks' – Chief Engineer Bill Miller checks fuel levels with Assistant Purser Allan Condie looking on, August 1976. (Second City Films)

On Sunday 4 July 1976, in conjunction with the Scottish Tourist Board, *Waverley* celebrated the bi-centenary of American Independence by operating a special public cruise. Stars adorned the funnels, the crew dressed up, and the Stars and Stripes flag was flown. (Second City Films)

Although the cruise started from Glasgow, it was the afternoon cruise from Ayr which attracted a large number of passengers. (Second City Films)

En route down to Ayr, a call was made at Dunoon where WSN made a presentation of a bottle of White Horse whisky and a model of the *Waverley* to the commander in charge of the US Navy Polaris base at the Holy Loch. Pictured, left to right, are: Terry Sylvester, Douglas McGowan and the US Navy Commander. (Second City Films)

The presentation drew a large crowd on the pier and the US Navy in turn presented WSN with a brass plaque which is still displayed today in *Waverley*'s dining saloon. From left to right: Bob Wilson (WSN), US Navy Commander, Douglas McGowan, a US Navy Captain, Captain David Neill (WSN), Jim Wallace (WSN) and Peter Reid (WSN).

WAVERLEY

☆☆☆☆☆☆☆☆☆☆☆☆☆☆☆☆☆ SPECIAL CELEBRATION LUNCHEON ☆☆☆☆☆☆☆☆☆☆☆☆☆☆☆☆☆

★★★★★★★★★★★★★★★ American Bi-Centenary Celebrations ★★★★★★★★★★★★★★★

☆☆☆☆☆☆☆☆☆☆☆☆☆☆☆☆☆ Sunday, 4th July, 1976. ☆☆☆☆☆☆☆☆☆☆☆☆☆☆☆☆☆

MENU

Silver Dollar Corn

or

Yankee Soup

******

Roast Turkey, Sweet Corn Pancakes & Bacon Chipalatos Boston Style

Vegetables	*Potatoes*
Stuffed Zucani *Creamed Artichokes*	*Baked Sweet Potatoes*

******

Blueberry Pie & Super Vanilla Ice

******

After Eight Mints

******

Wine sold by the glass £0.30.
(Red or White)

£2.00 inc. V.A.T.

WAVERLEY STEAM NAVIGATION CO. LTD.
IN ASSOCIATION WITH THE SCOTTISH TOURIST BOARD.

A special lunch menu was offered which quickly sold out; £2 seems astonishingly good value, even for 1976! (Douglas McGowan collection)

The ship's officers, photographed on the starboard landing platform at Tighnabruaich, September 1976. From left to right: Chief Officer Murray Paterson, Graham McLeod, Cameron Marshall, Derek Peters, Captain David Neill, Allan Condie, Archie McDonald, John McLeod and Neil McLeod. (Second City Films Ltd)

Waverley pursers, 1976 season. From left to right: Cameron Marshall, Fraser McHaffie, Derek Peters, Allan Condie and Graham McLeod. (Douglas McGowan)

Waverley's starboard paddlebox disappears into a trough off the Isle of Man on passage from Campbeltown to Liverpool. This was *Waverley*'s first foray south and at the time was thought to be a quite an adventure. In the light of what *Waverley* has achieved since, it was fairly tame! 29 April 1977. (Ian Shannon)

Opposite top: *Waverley* berthed at Princes Landing Stage, Liverpool, with the familiar outline of the Royal Liver building in the background. 30 April 1977. (Ian Shannon)

Opposite below: Alongside at Liverpool. The clock on the Royal Liver building shows 5.50 p.m. and *Waverley*, flags flying, is getting ready to receive the Lord Mayor of Liverpool with his entourage at a special open evening which included the general public. (Ian Shannon)

Holiday in Scotland
WAVERLEY
GLASGOW

Waverley departs Princes Landing Stage for a River Mersey cruise with a large party of Liverpool schoolchildren on board, May 1977. (Douglas McGowan collection)

Waverley, dressed overall for a special charter party, approaching Gourock in June 1976. MV *Queen of Scots* can be seen in the background, a vessel which was to play such an important role only a year later. (Ian Shannon)

Looking resplendent in fine weather, flags flying, *Waverley* makes her way up her native Loch Long. June 1977. (Ian Shannon)

Opposite: Backing away from Llandudno pier on 1 May 1997, Waverley was the star attraction in the pier's centenary celebrations. This event drew huge crowds not only on the pier but into the resort generally and was deemed a huge success by the organizers. (Ian Shannon)

Waverley has just turned hard to port as she swings into the north basin at Ayr Harbour. The engines have been rung full astern to bring her to a stop. Next, the heaving lines will be thrown; she will swing round on the cant rope, and will back across the harbour to the south wall. Early July 1977. (Ian Shannon)

Waverley berths at Kilmun pier in the Holy Loch on a special PSPS charter, September 1977. (Ian Shannon)

Tragedy strikes! *Waverley* lies pinned to the Gantock rocks on 15 July 1977. The pilot cutter *Gantock* (ironically!) stands by. (*The Scotsman*)

It's about 4.30 p.m., some two hours after *Waverley* grounded and Western Ferries' *Sound of Shuna* has successfully evacuated all *Waverley*'s passengers to Dunoon pier but is still standing by as the tide continues to recede. Two of *Waverley*'s lifeboats are in the water alongside the stricken paddler and the tug *Flying Demon* is preparing for another attempt to free *Waverley*. (Beaverbrook Newspapers)

The crowds at Dunoon's 'crazy golf' stare in utter disbelief at the surreal sight of *Waverley*'s thrashing paddles as she makes a final attempt to free herself from the rocks. (*The Scotsman*)

Waverley's stern begins to settle down on the Gantock rocks as the tide continues to ebb. All passengers have been safely landed without any casualties. (*The Scotsman*)

The tugs *Flying Mist* and *Flying Demon* get ready to take a line in another attempt to free *Waverley* as *Sound of Shuna* comes alongside. This was in itself a very difficult manoeuvre as *Sound of Shuna* faced a risk of grounding herself. (*The Herald* and *Evening Times* © SMG Newspapers)

Early morning at Dunoon's coal pier the following day, 16 July 1977. A forlorn *Waverley* sits on the bottom as the damage to the ship's hull is examined. (George Young Photographers)

A distraught Douglas McGowan on *Waverley*'s bridge, shortly after the stranding. This was shot by an enterprising *Scotsman* photographer who had dashed to Dunoon from his Glasgow office, hired a drive-yourself motor boat and clambered aboard *Waverley*, totally uninvited! (*The Scotsman*)

9.00 p.m. on Tuesday 19 July 1977 and *Waverley* survives her last low tide at Dunoon, sitting on the bottom. Additional pumps have been set up and one of the outlet hoses protrudes from one of the dining saloon windows. Temporary repairs have been carried out and the steamer is slowly raising steam. The next day she will cross to Greenock for dry-docking. (George Young Photographers)

Waverley steams past Greenock's esplanade at 17 knots from Dunoon to the Garvel dry dock at Greenock for a full examination following the grounding on the Gantocks. Note how low she is in the water. At this stage, no one knew if the ship would sail again. 20 July 1977. (George Young Photographers)

This photograph, taken in the Garvel dry dock, gives an impression of the considerable damage to *Waverley*'s hull following the Gantocks episode. The buckled section is where the ship settled on the rocks at low water. (George Young Photographers)

MV *Queen of Scots* at Millport, August 1977. This little vessel was used to maintain an amended timetable while the *Waverley* was out of service. Without that critical cash flow, it is unlikely the company would have survived. (Douglas McGowan)

Above: The officers and crew at Rothesay at the end of a traumatic season. Front row, left to right: Douglas McGowan, Roddy McIsaac, Allan Condie, Cameron Fernie, Chief Officer Murray Paterson, Captain David Neill, Ian Muir, Bill McFarlane, Angus McLean, Derek Peters and two members of the catering staff. September 1977. (George Young Photographers)

Right: Same occasion but the ship's officers only. Front row, left to right: Chief Officer Murray Paterson, Captain David Neill, Ian Muir (Chief Engineer) and Bill McFarlane (Chief Steward). Back row, left to right: Cameron Fernie (Second Mate), Derek Peters (Purser), Allan Condie (Junior Purser) and Angus McLean (Second Engineer). (George Young Photographers)

SATURDAY July 16 1977

Daily Record

7p SCOTLAND'S BIGGEST DAILY SALE No. 25,464

Mayday as Waverley smashes into reef

HIGH AND DRY!

THE trouble-hit Waverley ran into more trouble yesterday.

She crashed on to the Gantock Rocks in the Clyde near Dunoon.

But last night the Waverley—the world's last sea-going paddle steamer—was safely refloated.

Picture by ANDREW HOSIE

At high tide one of the tugs standing by pulled her off at the first attempt.

She then sailed a mile under her own power to tie up at Dunoon's coaling jetty.

After the vessel ran aground the 621 passengers on board took to the boats or were lifted off by rescue craft which dashed to the scene and stood by alongside.

The crew stayed on board . . . working flat out to stop water coming in three holes.

Now the vessel will go into dry dock

ABANDON SHIP—PAGE THREE

CITY EDITION

Scotland's Newspaper

GLASGOW HERALD

195th year—No. 146 SATURDAY, JULY 16, 1977 Eight pence (8p)

Tugs pull Waverley off rocks

Tourists cheer as steamer is refloated

By CARL GORDON, JIM HEWITSON, and CHARLES GILLIES

Kirn
Gourock
Dunoon
Cloch Lt.
Gantocks

The paddle steamer Waverley was refloated from the Gantock Rock off Dunoon shortly before midnight after a tense nine-hour battle in driving rain and against the tide.

Headlights from the cars of hundreds of holidaymakers illuminated the scene 400 yards out from the Dunoon pier. As two tugs eased the ship from the lighthouse rock a joyful cacophony of car horns could be heard for miles around.

A question mark had hung over the future of the Clyde paddle steamer after it ran aground during a Fair Friday "Doon the Watter" sail.

The ship's band had just finished playing "Amazing Grace" yesterday when the 30-year-old Waverley, the last sea-going paddle steamer in the world, sailed on to the treacherous Gantocks.

A day out in flat, calm conditions came to an abrupt halt for the 641 passengers and within minutes a fleet of ships was racing to the scene.

After a wait of about 40 minutes for the rescue ships, during which the passengers were moved about the ship en masse in an attempt to shift it, the order, "Women and children first" came over the loudspeaker.

The owners of the ship, the Waverley Steam Navigation Company, said last night the accident was caused by a breakdown in the steering gear.

The owners o fthe ship, a day-long £1.95-a-head cruise from Glasgow to Lochgoilhead, calling at Dunoon, is understood to be holed in two compartments and taking water.

The passengers, many of them women and children, were taken off unhurt by the Western Ferries car ferry Sound of Shuna, a pilot cutter, and the lifeboats.

The Clyde steamer Juno hove to for about an hour and two US Navy tugs, as well as the police launch Semper Vigilo and other small craft, remained nearby.

Cox John MacKinnon, aged 37, of the pilot Cutter Gantock, was at Gourock when alerted. He packed 84 passengers from the Waverley into his small craft.

He said later: "They were very orderly. I was surprised at the marvellous way they reacted. There was a lot of laughing and joking."

Passengers were put ashore at Dunoon and then taken to Gourock to resume their journey to Glasgow by train.

Mr William Smith, aged 72, who plays the concertina in the ship's band, said: "We were nearly thrown on our backs when the ship struck. We could hear the groan and scraping as she sat down on the rocks.

"But there was no panic and the people were composed. The engines were going full steam but the ship would not move."

Mr Joe McGroarty, a 68-year-old violinist, and Mr Bill Tennant, aged 71, the accordionist, said they had just packed up their instruments after playing "Amazing Grace" when the ship grounded.

The Clyde tugs Flying Demon and Flying Mist returned to the scene at 7.30 p.m. to prepare for another attempt to refloat the Waverley.

Evening low tide, when the Waverley faced a severe test with her keel stretched over the bare rocks, passed without incident.

The fear had been that she might break her back on the reef as the tide dropped but she still proved resilient. Hundreds of holidaymakers and local people packed the promenade as the hull of the vessel slowly emerged from the water.

Her stern was 12 or 14ft. clear of the water by the lowest point of the tide and the famous paddles were also clear of the surface.

A Clyde Port Authority cutter cruised around the vessel and, with the crew of the Waverley still on board, a lifeboat was moored to the paddle steamer's side in case of an emergency.

An attempt to free her within a couple of hours of her grounding failed but the tugs succeeded in manoeuvring her into what is hoped would be a better position for pulling her off at the next high tide.

Passengers who disembarked at Gourock and got off the trains at Glasgow Central station, some of them three hours late, were cheerful and none the worse for their adventure.

Douglas Bathie, aged 28, of Devon View Road, Airdrie, who was with his parents on the sail, said: "I knew something was going to happen when I saw the ship coming pretty close to a buoy. I was told to jump into a lifeboat which I did."

Dr Morag Bratten, of Lusset House, Old Kilpatrick, said: "We were told over the tannoy to keep away from the propellers after having heard a tremendous crunch.

"Some one shouted 'Women and children first' and at the time we thought it was a bit of a lark. We soon discovered it was not. Two lifeboats were lowered and some people went into them."

Mr Dougald Mitchell, aged 81, of Crown Avenue, Clydebank, said: "The boat tilted up and then came back again listing to starboard. There was no panic but a few people seemed to be upset. No one told us at the time what was happening.

"We were all told to move to one side of the ship and then to the other."

A 13-year-old girl, Delia Menzies, of Anderson Drive, Aberdeen, said: "A lot of people seemed to be afraid and one woman near me became ill. After about 40 minutes we were taken off by the Sound of Shuna.

"No one seemed to realise what had happened until the other boats started arriving around the Waverley."

A group of six students from Strathclyde University were also on board and one of them, 22-year-old John Hamer, of 79 Great George Street, Hillhead, became an oarsman on one of the lifeboats.

He said later: "They asked if there was anyone who could row and I volunteered. We must have rowed about 200 yards to the Juno."

The Ill-fated Clyde Legend
— Page 3.

Passengers disembark from the Sound of Shuna at Dunoon after being taken off the paddle steamer Waverley (right, background), aground on the Gantock Rock.

Crunch came after steering fault

Waverley director Douglas McGowan, of Stanmore Drive, Mount Florida, Glasgow, was on board the steamer as it grounded on the Gantocks.

He is a founder member of the Scottish branch of the Paddle Steamer Preservation Society and was a prime mover in saving the ship from the scrapyard.

He described last night the events leading up to the incident. "It appears at the moment that the ship did not answer the helm as it swept into Dunoon Bay. It was then put astern and grounded on the rocks.

"It is taking in water aft but at the moment the pumps are containing it.

"Obviously, the Waverley is insured and if we can get it to Greenock for dry-docking there is a possibility that it could be back in service within a week. We can only cross our fingers and hope."

LATE NEWS

HOME NEWS

JOHN EASTON recounts a chapter of accidents

Ill-fated paddle steamer that is a Clyde legend

Captain David Neill

The future of the Waverley is of vital concern not only to hundreds of private individuals who have invested in the ship as a unique tourist attraction, but also to local authorities and the Scottish Tourist Board, who have pumped thousands of pounds into the preservation effort in the last three years.

Ironically, yesterday was not the vessel's first encounter with the Gantocks, a group of rocks off Dunoon which 11 years ago claimed the lives of seven crewmen from the Swedish ship, Akka. It sank after foundering on the reef, which is marked by a large beacon.

In 1951 the 694-ton Waverley, then a railway steamer, was damaged near the same spot when it ran aground in thick fog, although it was later able to sail under its own power to Gourock.

The Waverley's career got off to a chequered start soon after its maiden voyage in 1947 when it was caught by the wind while reversing out of Arrochar Pier and turned on to the mussel bank at the head of Loch Long.

The vessel, the world's last sea-going paddle steamer, was built by A. and J. Inglis, Ltd., of Glasgow, and named after a Clyde steamer which was lost at Dunkirk.

It was withdrawn at the end of the 1973 season and the following year was sold by Caledonian MacBrayne for a nominal £1 to the Paddle Steamer Preservation Society, who later formed the Waverley Steam Navigation Company.

How the steamer came to grief yesterday on a notorious landmark will automatically be the subject of a Board of Trade inquiry.

It was in charge of an experienced Master, Captain David Neill, formerly employed by Western Ferries and was on a regular Friday sailing from Glasgow to Dunoon and Loch Goil. One theory was that its steering gear failed.

The Waverley story has captured the public imagination since it was handed over to enthusiasts who mounted a major campaign to keep the vessel in operation.

Strathclyde region withdrew its grant after the first year and it was left to district councils, including Glasgow, Kyle and Carrick, and Cunninghame to contribute along with the STB and private subscribers to ensure that it would continue to give pleasure to thousands on the Clyde.

Apart from Captain Neill, himself a director of the Waverley Steam Navigation Company, the main driving forces have been Mr Terry Sylvester, owner of a chain of do-it-yourself shops in his native Wales, and Mr Douglas McGowan, a chocolate salesman from Mount Florida, Glasgow.

With others, they parted with large sums of their own funds as a gesture of faith in the future of the ship.

Expensive repairs have been carried out, including re-tubing of the boiler—itself costing some £40,000—and this year the passenger accommodation has been refurbished with aid from Tennant Caledonian Breweries and the White Horse Distillery Company.

Since the take-over by the private group there have been frequent breakdowns because of smashed paddle floats.

On one occasion passengers had to disembark in small boats off Tarbert, and last summer the Waverley had to be towed off the beach at Rothesay.

But until yesterday it was reckoned that most of the difficulties had been resolved and that the steamer was set fair for a rosy future.

In May the Waverley broke new ground by sailing to Liverpool and Llandudno and its owners were delighted with her performance in a fierce gale.

Before yesterday's grounding it had already carried more than 100,000 passengers this season. The vessel's owners remained optimistic that it could be back in service in a few days.

A view from the air of the Waverley, with her lifeboats aft, on the Gantock Rock. The car ferry Sound of Shuna is in the foreground.

The Sound of Shuna alongside the Waverley.

Passengers from the Waverley . . . (from left) Dr Morag Bratten, Catrina Bratton, and Delia Menzies, last night at the Central Station, Glasgow.

EVENING Times

SCOTLAND'S GREATEST EVENING PAPER

No. 31,820 Saturday, July 16, 1977 8p

A FAIR BONANZA

● **It's a give-away Fair in your holiday Times**

★ There's a host of great prizes in the Times-Lewis's Fair Snips competition on Page 7.

★ *You could win a fabulous fortnight's holiday for two in Maderia or Tenerife. See Page 11.*

★ From Monday there will be EIGHT tenners to be won every day in our popular Match the Money contest—See Page 5 for today's lucky numbers.

WAVERLEY'S FUTURE IS NOW IN DOUBT

● Fairy lights brighten up the scene as the Waverley is finally berthed at the Coal Pier, Dunoon, early today after being towed off the Gantocks.

THE FUTURE of the paddle steamer Waverley hangs in the balance today as she lies damaged at Dunoon after crashing on to the Gantock rocks.

Words DAVID STEELE
Pictures JAMES MILLAR

And as the job of putting the Clyde's last paddle steamer back into service began, a question mark hung over the cause of yesterday's accident.

The Waverley was about to dock at Dunoon late yesterday afternoon during one of her pleasure cruises to Loch Goil when she ran aground on the Gantocks about 400 yards from the pier.

First reports indicated the steering gear has failed, but this will be the subject of a full investigation.

The operation to rescue the 621 passengers and refloat the ship took nine hours — much of it in darkness and torrential rain—and involved tugs, car ferries, and a flotilla of small craft.

The in-shore lifeboat from Largs stood by as the operation got under way.

The first task today for Board of Trade experts and the paddler's owners, Waverley Steam Navigation Company, is to decide whether the ship is fit to sail to dry dock—probably at Greenock.

First examination of the Waverley's hull as she lay in low water at the Coal Pier showed there are at least four punctures in the afte section and a severe bend in the hull.

SEALED

A decision will be made later today as to whether and when the Waverley can be moved and it is expected that quick-setting concrete will be poured into the lower hull to make a temporary seal until the ship can be dry-docked.

Fair holidaymakers lined the promenade watching the delicate refloating operation until the early hours of the morning.

As well as damage to the hull and one of her two tall funnels—bent by the impact of the grounding—the decks and cabins are strewn with debris.

Many windows had to be broken to let out exhaust fumes from the pumps working inside and the remains of meals and drinks which the passengers were taking when the Waverley went on to the rocks still lie on the cabin tables.

Lifejackets worn during the rescue are strewn around the decks and the ship's lifeboats which were launched last night are still hanging down the side of the Waverley.

Among the heroes of last night's operation were a warrant officer and 20 ratings from the US nuclear base in the nearby Holy Loch.

They worked throughout the night with the crew using powerful pumps to get out the thousands of gallons of water which had flooded the lower decks.

Douglas McGowan, the Glasgow man who is director of the Waverley Company, said — "It is not certain when we can get her back into service. There is a lot of work to be done."

● High and dry . . . the paddle steamer's damaged hull (arrowed).

WEATHERCAST

unny periods; scattered showers; ow this evening; cool, fresh wind.
ighting-up time — 10.20 p.m.

A MAN AND HIS DREAM

Douglas McGowan and the Waverley . . . "She'll keep coming back, she'll never die."

Story: JAMES McBETH
Pictures: RICHARD PARKER

'I won't let Waverley go under'

DOUGLAS McGOWAN wandered through a Clydeside dock yesterday, his dream shattered.

For Douglas is the man who fought to save the paddle steamer Waverley from the breakers yard.

He survived crisis after crisis, only to be hit by the biggest disaster—when the ship went aground last Friday.

Now she is sitting in Scott Lithgow's Garvel yard in Greenock waiting for the four large gashes in her underbelly to be repaired.

Yesterday, Douglas went to inspect the damage and despite the £40,000 cost of repairing her, he is determined that his dream won't die.

He vowed: "The Waverley will be back in a month. We'll find the money somehow."

REPAIRS

The ship can stay in the dock only for a few days. Scott-Lithgow's need her berth for a submarine due in next week.

Douglas, a member of the Paddle Steamer Preservation Society and a director of the Waverley Steam Navigation Company, said:

"We'll have to find another dry-dock near Glasgow but we'll manage.

"This yard will do emergency repairs for us, installing steel girders along part of the hull."

Douglas inspects the gashes in the steamer's hull.

Life hasn't been easy for the Waverley since she became a financial embarrassment to Caledonian-MacBrayne in the early 70s.

She was all set for her last cruise to the breakers yard but the PSPS fought to save her until Calmac agreed to "sell" her for £1.

Douglas was the man who handed over the society's cheque.

After the sale it took a year and £73,000 to bring her back to life as the pride of the Clyde.

Glasgow threw in £6000, £1000 came from a city couple and an OAP even chipped in her 50p.

PROMISE

Grants and donations kept her going and this year, with the promise of £40,000 in advance bookings, her future seemed assured.

Then came the disastrous crash on to the rocks off Dunoon.

Daily Record

Before the start of the 1978 sailing season, Waverley Steam Navigation Co. was approached by the BBC to ask if they could use *Waverley* for a live broadcast of the *Noel Edmonds Show*. Agreement was reached and several other well-known disc jockeys were in attendance, including Tony Blackburn, pictured here with two admirers on 13 March 1978. (*The Herald* and *Evening Times* © SMG Newspapers)

Noel Edmonds broadcasting live to the nation from *Waverley*'s dining saloon. The steamer played host to an audience on board of over 1,000, at 7.00 a.m. in the morning! 13 March 1978. (*The Herald* and *Evening Times* © SMG Newspapers)

A number of weddings and receptions have been held aboard *Waverley*, although the ship is not officially licensed for the actual legal ceremony. None were more spectacular, catching the front page of a number of national newspapers, than that of Bill and Sandra Purves, pictured here at *Waverley*'s stern on the ship's first-ever visit to London, 29 April 1978. (*The Herald* and *Evening Times* © SMG Newspapers)

Well-known personalities are frequently to be seen treading *Waverley*'s decks but none as dedicated or enthusiastic as the actor Timothy West and his actress wife Prunella Scales, seen here in *Waverley*'s wheelhouse with Andy Dodds on the helm. (Douglas McGowan)

London's impressive Tower Bridge lifts for *Waverley*'s first excursion from Tower Pier. This was yet another poignant moment in the early years of preservation: *Waverley* had conquered the capital! April 1978. (John Goss)

Waverley berthed at Poole harbour. This is not a particularly easy harbour for *Waverley* to navigate and she has made only a few visits since May 1978. (Ian Shannon)

Arriving at Bournemouth for the first time, May 1978. (Ian Shannon)

Berthing in Torquay harbour, May 1979. Weather conditions on this occasion were difficult, with winds gusting to Force 6. Both bow and stern ropes are secure but the strong winds have blown *Waverley* away from the quay wall. The anchor has been dropped to steady her until the windlass and capstan can pull her alongside. (John Goss)

Waverley swings to port past Sun Pier, Chatham, into Limehouse Reach on the River Medway. Historically on this cruise each year, *Waverley* has met up with the other PSPS paddle steamer *Kingswear Castle* in a 'Parade of Steam'. 13 May 1979. (John Goss).

Leaving Worthing pier, May 1979. This is a very exposed location with a frequent swell and conditions have to be fairly calm before *Waverley* is able to call here. (Ian Shannon)

Waverley departs Newhaven harbour for Portsmouth, 10 May 1978.

Two Clyde-built passenger ships – little and large! *Waverley* passes the *QE2* at Southampton, May 1978. (Ian Shannon)

Passengers embark at Greenwich, 3 May 1978. This pier is no longer a regular call for *Waverley*. (John Goss)

Alongside Tilbury landing stage, 28 April 1978. *Waverley* is getting ready to embark her passengers for a Radio Medway charter to Sheerness and Rochester. (John Goss)

Arriving at Gourock to pick up a special charter party, May 1978. (Ian Shannon)

Officers and crew, pictured at Stobcross Quay at the end of the 1978 season. General Manager Kyle McKay is fourth from the left, front row. September 1978. (George Young Photographers)

Opposite top: Berthed at Kilmun pier with a good complement of passengers on a special PSPS charter sailing, September 1979. (Ian Shannon)

Time to hand over: Douglas McGowan wishes every success to Jim Buchanan, a Helensburgh businessman, in February 1979. (George Young Photographers)

Twenty one years later! John Whittle and Douglas McGowan meet up again at the PSPS Scottish Branch Dinner at Glasgow's Art Club, 18 November 1994. Douglas presented John with a framed print of *Waverley*. (Douglas McGowan collection)

Waverley arrived back in Glasgow from stage one of her major rebuild programme in Great Yarmouth on Friday 18 August 2000. Her captain, Graeme Gellatly, and a Scots piper wish her good fortune before the inaugural evening cruise. (Douglas McGowan)

Paddling by moonlight. A late evening photograph of *Waverley* returning to Newhaven harbour following an evening cruise, 22 April 1978. (John Goss)